Alien Intentions

By John Buckner

This book is a work of fiction. Names, characters, businesses, organizations, places and events are either products of the author's imagination or are used fictitiously. Any resemblance to actual persons, living or dead is purely coincidental.

Chapter 1

It was one of those lovely spring days when everything seemed to be fresh and clean and the air crisp and invigorating. The two couples were strolling across the campus of the University of California at Berkley after their final class on Friday afternoon verbally wondering how they were going to spend the promising weekend. The weather forecast was for the same type of weather for the next three days and none of the four relished spending the time on campus.

Eddie Casteen was a jock, or at least liked to think of himself in those terms. He was not much for the team sports but enjoyed running and had become a very good competitive runner, especially in the endurance events. He was not a sprinter, wasn't even particularly fast, but he had the stamina and determination to push himself beyond where most runners give up. Anything longer than the mile he showed well no matter the caliber of the competition. He was just short of six feet tall and slender. He only weighed one sixty five and his body fat was almost microscopic.

Beth Glover was his girlfriend and had been since they had met two years previously while registering for their freshman classes. Something had sparked immediately and they had their first date the following evening. It was well that they had a lot of the same interests because Eddie spent an awful lot of time training. He did not consider it training as he was doing something he loved to do and would have done it even if he wasn't on the track team. Beth was a runner too, although her running was to keep her trim figure and a way to clear her mind. She always seemed to think better when she was on the move, her circulatory system pumping, and the breath coming in labored gasps. Beth was a very pretty young lady. The guys said she was hot, which was a high compliment indeed. Her blond hair and green eyes gave her a piquant look that would go well in Hollywood.

The other couple was Bill Jenkins and Cindy Norman. Bill was five feet ten inches tall and weighed just short of two hundred pounds. He was heavy, but not fat. He didn't overeat and seldom ate sweets. He just had a very low metabolism and it took him a long time to burn the calories he ingested. Bill's hair was blond and his eyes a pale blue. He wasn't a jock in the sense that he went out for sports, but he was well coordinated and could play most team sports well enough not to be embarrassed by his performance. He spent more time with books than most his age and seemed to have an insatiable zest for knowledge.

Cindy Norman had dark auburn hair, which she usually wore in a pony tail, although that style was not very popular at the present time. Her eyes were gray, which was an odd color considering her hair. Cindy too was a good looking girl. She was an aspiring actress, or wanted to be after she finished school. She participated in all the school plays, which is where she and Bill had met. They were both trying out for parts in a school play and had teamed up to read lines for the audition. Bill was reading the part for the male lead and Cindy for the female lead. Their chemistry was such that the director chose both for their respective parts. They were together a lot for rehearsals and performances and the fact that they were romantically involved in the play, only made it seem natural to carry it on into real life.

Beth and Cindy had somehow ended up as roommates because neither of them had listed a preference for roommates on their applications. Both were pretty, intelligent, and more mature than the run of the mill freshmen, at least that's how they saw themselves.

As they strolled across the campus toward the dormitory Bill said, "I can't see wasting a lovely weekend like we have coming up cooped up here, can you guys?"

Beth said, "What do you have in mind?"

"I don't know, just something away from here."

Cindy said, "Why don't we go for a drive in the mountains?"

"Which mountains do you have in mind?" asked Bill.

"I don't know, maybe the Sierras. It will be a lot colder up there. Maybe we can take the camping gear and sleep out tomorrow and Saturday nights," Cindy said.

"Let's do it. It sounds like fun and we need to recharge the batteries so to speak. You guys didn't forget that we have finals coming up in two weeks?" Beth asked.

"Then we take our books, though if we are going to hike to a campsite I am not taking the extra weight," Beth said.

"Didn't you learn that the guys are always supposed to carry the girls books?" asked Cindy.

"How could I forget something that important?"

"If we are going to do this we had better get a move on," Eddie said. "Let's go to the dorm and get our sleeping bags and books if you guys insist. We can stop on the way and stock up on food and munchies. The tents are in my truck." It would not occur to any of them until much later that their fate had been decided that quickly. They went to the dormitory and packed small bags, grabbed their sleeping bags and headed to Eddie's truck, which was a crew cab. They transferred the tents to the bed of the truck with the rest of the gear. Eddie always carried a medium sized cooler, in which he kept stowed basic cooking utensils, a butane lighter and salt and pepper.

As they headed out of town they stopped at a shopping center grocery store. Beth, Cindy and Bill did the grocery shopping while Eddie got gas for the truck and ice for the cooler. He also picked up a map put out by the forestry service which showed the forest roads and trails in the area around Yosemite. He had not consulted the others but thought they would like the country up that way.

When they all met back at the truck in the parking lot they reviewed their purchases to make sure they wouldn't be out in the middle of nowhere and suddenly realize that they had forgotten the toilet paper. Eddie looked in his tool box to make sure he had the hammer and hatchet that he usually carried. Satisfied that they had what they needed Eddie got them back on the road.

Even leaving as early in the day as they had, they were still quite late getting to the area Eddie had chosen. It was already almost dark when Eddie pulled off the rough forest service road he had taken to get them away from the better traveled areas. He found a meadow which they could barely make out in the fading light. Eddie horsed the four wheel drive truck to the far side of the meadow and found a fairly level spot under some trees. Even though he knew that it was illegal to camp in non designated areas he figured the area was remote enough that anyone discovering them would be a fluke.

Eddie parked the truck so the lights would illuminate the area well enough for them to be able to erect the tents. There was only one flashlight and they wanted to save the batteries to look for firewood. They were above the eight thousand foot level Eddie estimated and the temperature was already colder than in Berkley and would become much colder as the night wore on.

It only took them a few minutes to erect the tents and the men went to gather firewood while the women planned the meal, roasted hot dogs and potato chips.

Within half an hour they had a cozy fire going in a fireplace the women had fashioned by arranging rocks in a three foot diameter circle. There was an ample supply of rocks, many of which they had removed from the tent floor to make the sleeping a bit more comfortable.

Soon they had the hot dogs roasting on straightened coat hangers. They had not had supper and the hot dogs smelled good to all of them. As they sat around the fire eating their hot dogs and talking all conversation suddenly came to an end, almost as if they had exhausted the topic and no one wanted to say anything more. The verbal conversation ceased, but something more sublime was taking place. Each of them seemed to feel that the verbal conversation was continuing yet was somehow different.

A voice was addressing them collectively but the source of the voice could not be seen. It had a dreamlike quality and seemed to be familiar to each of them. It said, "Welcome, you are my first visitors for almost three thousand of your earth years."

Bill said, or thought he said, "Who are you? And what are you?"

The voice replied, "My name does not matter and wouldn't mean anything to you. You can just call me Joe if you wish."

"Am I dreaming, or are all of you witnessing this?" asked Eddie.

"I don't think you are dreaming and from the expressions on your faces I think we are experiencing the same thing, whatever it is," Cindy said.

Joe said, "I don't mean to alarm you, but I have not had contact with your kind for so long that I had to reprogram my circuits. Let me explain myself a bit more clearly. It will be easier if I bring you to my presence. Will you allow me to do that?"

"What if we say no?" asked Bill.

"I would not want to do anything against your wishes, but I think you are all curious enough to want to find out what this is about. I could bring you against your wishes, but that is not the proper way to start a relationship."

"What do you mean relationship?" asked Eddie.

The girls had been silent through these exchanges. "I think he means that he is powerful enough that if he wanted to he could have just whisked us off to wherever and we wouldn't have had any say in the matter. Is that about it Joe?" Beth asked.

"Well, physically I would have a hard time doing that, but your minds are the main thing that interests me."

"Interests you in what way?" Cindy asked.

"That's precisely what we need to discuss. I want to lay out all the facts and give you some background so that you will understand what I am going to tell you and ask of you."

"What if we agree to the visit, but don't like what you have to say?" Bill asked.

"Then you return here and I will erase this meeting from your minds and you will not remember anything about it."

"One question before we make a decision. You are not from this planet, are you?"

"No, but I feel almost like a human I have been here for so long."

"How long have you been here?"

"Almost ten thousand years in the way you measure time."

"Okay, let's go to your place. I assume you guys agree?" Eddie asked the others.

Suddenly they were in an area filled by bright light and indistinct shapes. One of the shapes beckoned them toward an area off to the side that had more subdued lighting. They moved to that area and the form of a man came with them. While his form was human in shape his head was more rounded than normal and his limbs seemed to have more joints. The arms had three joints as opposed to two, and the legs the same. The hands had five fingers with thumbs on both sides of the hand.

"I come from a world at the far side of your galaxy. My species was bred to be the peacekeepers of the universe. We are prepared for our tasks and transported to the assigned location where we remain until our relief arrives. We are not organic in nature and therefore do not require any sustenance such as food and water. We only require some form of energy to power our systems and to recharge our own beings at times. My task is to keep your world on track in terms of its evolution in technical developments and assure your survival until such time as you are capable of surviving outside your earthly confines. I am tasked to do these things without direct intervention; hence I must contact beings such as you through which to work. I can do many things through your mental capacities. Your species only use a small portion of your mental capability and little more than ten percent of your physical ability. I can increase your mental ability to a degree that you will be more intelligent than anyone who has lived on your planet and your physical ability such that you will be more capable than most of your kind."

"And why would you do those things?" Bill asked.

"To save the earth from sure destruction."

"How is the earth going to be destroyed, and when?" Eddie asked.

"I cannot reveal that to you as it will influence your future actions, and that is not allowed."

"Not allowed by whom?"

"By my processors."

"Back where you come from?"

"Yes. We have very specific procedures to follow to prevent us from tampering with something that could affect the future of the universe, not just this planet."

"If we agree to save the earth, in your words, what do we have to do for you?"

"Not a thing. I will do what is necessary to prepare you to do what is required when it is required."

"Can you tell me how you know that the world is going to be destroyed?"

"Not in any way that you would understand, but maybe in a way that you can relate to. Have you read your Christian Bible that is so prevalent in your society today?"

"Some, but what has that got to do with the destruction of the earth?"

"Read the last book of the Bible and you will learn of the prediction that God will destroy the earth in His time."

"That is no different than me telling you that I predict the earth will be destroyed, only in the manner of the individual's belief system. If you are a Christian then you believe that God is going to destroy the earth, after taking His believers away first. I simply tell you that the earth will be destroyed without taking anyone away first."

"But with your help, we will be able to prevent this catastrophe?"

"Maybe, then again maybe not. It depends on what you do with the tools I provide you."

"So this isn't a sure thing no matter what we decide. If we agree to help, we don't get any clues to point us in the right direction, and we have to figure this out for ourselves to find out when or how the world is going to be destroyed, then take the necessary action to prevent it?"

"By George, I think you've got it. I heard someone say that in a movie once."

Beth said, "This just isn't making any sense to me. I think I heard you say that you were going to give us superpowers, both mental and physical, that would allow us to do almost anything we wanted. But you want us to use the power to stop some unknown person/nation/entity from destroying the earth next year/decade/century."

"It doesn't make sense to you now because you don't have the mental capacity to deal with it. That will change."

"I still don't get all this. Physics teaches us that opposites attract. And that for every action there is a reaction. Carrying this thought a bit further, black and white, night and day, hot and cold, good and evil... Do you see where I'm going?" Eddie asked.

"You are more perceptive than I thought. Yes I am the good guy and yes there is an evil twin, but he is bound by the same constraints as I am. He cannot directly intervene in any earthly matter, but he can approach people as I have approached you and try to convince them to destroy the earth rather than save it."

"So we might be going up against someone who will be our equal in the mental and physical departments."

"I don't think so. He is not as well trained as I am, and is mentally inferior."

"Do I detect a bit of vanity?"

"As I said you are more perceptive than I would have thought. I can only tell you that anyone he can convince to come up against you will be an individual, and as such cannot compete with the power of four."

"If we agree to this what will happen to us; I mean will there be anything tangible that we can point to and say, 'this is a result of our encounter with Joe'?"

"You will not feel any differently but you will experience a quantum leap in your mental acuity."

"And you won't be able to access our minds after this?"

"Only when you are in this precise location."

"If we agree to do this what's in it for us, other than saving the world of course?"

"Whatever you want. If you want to be rich you can make money in a variety of ways. Your expanded mental ability will enable you to see patterns that others don't. Take your stock market for example; you could invest in stocks that you know are going to rise. Real estate trends, and what people will buy are other examples. You might design gadgets that are important to your society. If you look at the scientific field there are new remedies you could design for control of disease, or to cure diseases."

"You're saying we will be capable of doing all those things?"

"You will only be slowed by your lack of ambition or interest."

"What do we have to do to accept your offer?"

"Put these devices on your head and I will take care of the rest. You will not feel anything at all."

Although they perceived that they were in a virtual reality world they could feel the texture of the helmet like devices he handed each of them. Cindy was feeling a little apprehensive but went along with the rest and placed the gadget over her head. Joe didn't seem to be doing anything and within a matter of seconds told them that they were finished. He told them, "Okay, we are finished here. Remember, you will only be able to contact me from this very spot. I will implant the geographical coordinates in each of your minds so that you will always remember."

"One last question," said Bill. "Will this make us immune to any of the calamities in our world such as disease, aging, or accidents?"

"I'm afraid not. You will still be subject to everything that your fellow beings are. The exception might be something intangible like being able to avoid an accident due to your enhanced physical prowess. Because your bodies will be more efficient you might avoid some of the more common maladies such as cold and flu because your bodies will fight the germs that cause them more efficiently."

"We are finished for now. I do hope that you live up to my expectations for you."

Without further ado they found themselves sitting around the campfire again. They looked at each other and Beth said, "Did I just have a dream about aliens or were you guys having the same dream?"

Eddie said, "Do you feel any smarter or stronger?"

"You did experience the same thing then?"

Cindy added, "That was really weird. Did all of you get the same message about saving the world from destruction?"

"I believe we did," Bill said.

"And what about becoming the smartest people on earth?" Beth asked.

"There's a way we might be able to test that theory," Bill said. "Cindy, grab a book and open it to a random page then read the first paragraph. Hand the book around and we will each read a paragraph."

Cindy did as asked and passed the book on. She was aware of reading much faster than was normal for her. She usually read slowly, especially when it involved study material. Each of them experienced much the same thing. When they had all read a paragraph Bill said, "Now Cindy, repeat what you just read."

She looked askance at him as if he was kidding her, but then repeated back verbatim the paragraph she had just read. Bill pointed to Beth and she did the same, as did Eddie and then Bill.

"Well, that settles that. We were not dreaming and have just had an encounter with an ET, though he didn't look like an ET. Let's see if we all had the same perception of Joe. Do you remember him or it telling us to call him Joe?"

They all nodded in the affirmative. "And do you remember his short lecture on how to make money?"

They nodded again. "Do you remember what he said about the evil twin?"

Eddie said, "I think that was just a metaphor, but there is evidently something or someone out there that means to destroy our planet, though Joe said he is not as powerful as he is, and that he can only influence a single person at a time, while Joe has done that with four of us. Is that what you guys understood?"

Cindy said, "He said that one was no match for four. That's assuming we all manage to live until the time comes for the confrontation or whatever is to happen."

"Well, it's 2015 now and we are all at about the twenty year mark. Average life expectancy is going to be near eighty years by the time we reach that plateau. That means that whatever he programmed us for is going to occur within the next sixty years," said Bill.

Beth added, "Assuming we don't get run over by a car, die in a plane crash, or succumb to some incurable disease. The last part is not as likely according to Joe because we can invent a cure for the disease if we are so inclined."

Eddie said, "If what we just experienced really happened, and I think we all agree that it must have, then our school days are over. We all seem to have photographic memories and anything we read will stick, so why bother about school. We go to school to learn an occupation in order to have a decent life style and we no longer have to worry about that. I believe we should spend this weekend developing a game plan for dealing with our situation."

They all seemed to agree with Eddie. Beth voiced what all of them were thinking. "If what Joe said is true, and it surely must be, then we will not have any worries about money or material things. That kind of takes away the motivation factor. Most of what everyday people do is geared toward fulfilling basic needs. We are going to have to set some long range objectives and figure out how to deal with the destruction of the world part of the equation."

"I'm wondering how to go about that without any clue as to what to look for. If there was some event or precursor it would certainly make the task easier," Cindy said. "I think if we approach the problem logically we can design a worksheet of building blocks. In order to destroy the earth there are only so many ways to do it based on today's technology. You have nuclear weapons and biological agents. If we look at those two things and determine what would have to lead up to an event of that nature we might be able to establish a rough framework of things to look out for," Bill said.

"What about the good old fashioned global conflict? Didn't Hitler almost succeed in his plan to wipe out the entire Jewish population? If one man could do that then another could do the same with regard to any other ethnic group and the rest of the world would have to choose sides."

"The third option would be the use of some weapon of mass destruction that hasn't been invented yet. Some weapon that shows up in someone's fantasy in a movie or book might just be turned into reality. Yesterday I would have laughed at someone who thought an anti-matter weapon was possible. Now I think it not only possible, but feasible in the not too distant future," Eddie said.

"Let's sleep on it and talk about it in the light of day tomorrow," Cindy yawned.

They banked the fire and crawled into the tents.

Chapter 2

When the two couples awoke on Saturday morning the sun was up but there was a chill in the air. The temperature was in the thirties and Eddie quickly gathered enough wood to get the fire going again. He put some water in a sauce pan to heat for instant coffee, while Beth broke out the skillet for the bacon and eggs. The meadow where they were camped was more picturesque in the daylight than it had been the night before.

"It really is beautiful here, isn't it?" Beth asked.

"Yeah, we couldn't have picked a more perfect spot even in the daylight," Eddie replied.

"Now that we are in the light of day, do you still remember everything about last night?"

"You mean what you did to me after we went to bed?"

"No silly, the meeting with the alien."

"I still remember it vividly. I want to test the physical part of what Joe said." He then picked up a good sized rock and threw it hard toward the woods. The rock went much farther than he was capable of throwing it the previous day.

"What was that about?"

"Just testing the physical part of what he told us. I guess that's true too."

Cindy and Bill joined them about the time the water was hot enough for the coffee. "Wow, what a night! I still don't know if we were all hallucinating."

"Bill, what is the most important thing you have to do in your lifetime?"

"Save the world from destruction?"

"As of last night I would have to say that's the highest priority. The two things we agreed would be the most likely based on today's technology were nuclear and biological. Suppose Beth and I concentrate on the nuclear aspect and you and Cindy take the biological. Your job will be much harder than ours since there are many more possibilities so we will give you some help. Beth and I will also concentrate on making money until we get better established."

"Are we agreed that school would be a waste of time for us now?"

A chorus of assent followed. "I wonder," said Cindy, "if we have children will they be like us in terms of intelligence, or will they just be normal?"

"I don't know, but it's a little too early to think about that just yet," Bill said.

Eddie said, "I believe the best way to get our finances off the ground is through the stock market, but it takes capital to buy stock. I have been thinking about a concept that could make a lot of money, but it will take some time to put it together. The gadgets he used on us last night got me to thinking about a thought computer. If we designed something with electrodes to attach to the head to read thought patterns and translate them to computer language that would be a totally new innovation. Also the computer can be operated hands free with capability limited only by the speed with which thoughts come to the user. It has been said that the mind is the ultimate computer. Instead of using digits to represent values why not use electrical impulses as the computer language? I will have to learn a lot more about the way the brain functions to sort it out, but I believe the concept is a good one. If it could be worked out it would bring about a complete revolution in the computing industry. Digital computers would become obsolete overnight, and we would be richer than Bill Gates."

"The patent alone would be worth billions if we were to apply for a patent, but I don't think that will be necessary because no one else will be able to build it since it will operate on neurological impulses and you can't harness those. I don't know if this will work out, but the logic of the concept is sound and there has to be a way to do it."

"You might be onto something, but it is probably a long way down the road. Our immediate concern has to be the here and now. I know none of us have any money to speak of. If we tapped our parents we might come up with ten thousand dollars. The question is how do we parlay ten thousand into enough to get started in the right direction?" Beth asked.

"If we picked a stock such as an IPO and could be assured that the company would succeed then we could compound the money pretty quickly. One way to do that would be to go to work for the company and ensure that it succeeded."

"I like that idea but it would be better if we could do the company start up and get a couple of patents for medicines then go public. Since I am interested in biology and the sciences maybe I can come up with something the science community has been puzzling over for a long time and use that as a vehicle," Bill said.

Cindy threw cold water on the conversation. "It's nice to sit here and theorize, but we don't know the reality yet of what we were told. Until we can test the precepts of a new way of life we need to figure out the limits and limitations. I've heard people say that a person is only limited by the scope of their imagination, but we are going way beyond that."

"I think it would be a good idea to stay in school for the rest of this semester. The housing is cheap and that will give us time to research the ideas that we come up with. The course work will definitely not present a problem," Beth said.

"You're probably right about that. I want to finish the track season anyway," Eddie agreed.

They spent the rest of Saturday and all day Sunday talking over possible ways to use their abilities. The consensus was that that they needed seed money to get started in anything with the potential to make a lot of money.

They were out of the mountains when Bill's cell phone rang. When he answered it was his mother.

"Where have you been all weekend? I have been trying to reach you for the past two days."

"I have been camping with friends in Yosemite. There is no cell service in the mountains. We're on our way back to school now. What is so important?"

"Your aunt Nell died last week and her lawyer called to tell me that we have been left a considerable sum of money in her will. Your part will be about fifty thousand dollars and he wanted to know where to send the check."

"When did she die?" Bill asked, knowing before she told him what she was going to say.

"Just after ten o'clock Friday night."

Bill was stunned, even though he knew before she told him. "Give them the address at school please mom."

"I did that already. I just wanted you to know about her dying. We were never that close and I can't figure why she left us so much money. I didn't even realize she had that kind of money."

"What was the total?"

"Two hundred thousand dollars."

"And fifty thousand was to go directly to me?"

"That's what the lawyer said."

"Well, thanks for calling mom. I will talk to you later." As he killed the phone he looked at the others. "I have just inherited fifty thousand dollars from an aunt that I was never very close to. Do any of you think this is a coincidence?"

"It sure came at an opportune time if it was," Cindy said.

"If this means what I think it means what does that say about Joe's contention that he couldn't influence events directly?"

"I would say it makes him a liar. There's a way to check though. Find out the date of the will. If it is earlier then we have no proof, but if it is dated Friday I think we need to reevaluate our situation," Eddie said.

"I don't think it is going to prove anything one way or the other. He would seem to be capable of orchestrating events well in advance," Beth added.

"We are in way over our heads. If what he did to us in such a short time is possible then he may be capable of much more. One of our problems is that we have no idea what kind of entity he is. He may just exist in the ether as plasma or some element or form that we are not familiar with. He seemed sincere in what he told us, but he could just be an accomplished liar."

"I intend to check the date on the will anyway. It is possible that he could somehow affect the past, but I don't know how."

"No matter, we have to play the hand we were dealt from this point on. There is no way to go back and change things. We just have to be sure of our motivations more than ever now," Eddie said.

They got back to their rooms before dark, still puzzled by the latest turn of events. None of them slept very well Sunday night thinking about the implications of Bill's aunt to their encounter over the weekend. Their new found intelligence caused them to lean more toward the death being a planned event rather than chance. The probability of the event not being connected was very low, but a being with the obvious intelligence of Joe would know that they would be able to see the connection. And wouldn't that work against what he had told them he wanted them to do? It was a conundrum that none of them could figure out.

On Monday Bill called his mom and asked her for the phone number of the lawyer who had called about the will. She gave him the number and he placed the call to the lawyer as soon as he hung up with his mom.

"Mr. Wilkins?" he asked when the phone was answered.

"This is Mr. Wilkins office. Who may I say is calling?"

"My name is Bill Jenkins. My mother got a call about my aunts will and I had a question about the will."

"Oh yes we called your mother on Saturday. What was the question? Maybe I can help you. I have the will here on my desk."

"I wondered about the date of the will. I had not seen my aunt for a long time and wondered when she made the decision to put me in her will."

"The date on the will is February 14, 2014. Apparently she thought to leave you a Valentine's present."

"Well, thank you for the information. You have been very helpful."

"You're quite welcome. You seem like such a nice boy," she said as she smiled to herself.

After she hung the phone up she glanced once more to the open office door of Dave Wilkins. She then picked up her purse from the desk and walked out the front door allowing it to lock behind her.

The two couples had breakfast together in the school cafeteria. Bill told them of the phone call he had made and the result of the call. "The will was dated February 14, 2014 according to the secretary. That would have taken some long range planning to set up such a trivial thing as start up funds. I mean, there are so many things that would have to be known in advance it just doesn't seem possible that it could be done."

"Let's just keep a close eye out for more anomalies. If it is true that Joe can't directly act to influence our actions then whatever we do will be of our own volition. I have a creepy feeling about this whole thing," Cindy reiterated once more.

"Well, how are we going to use the money? Do we look for a promising start up and help it along, or do we do our own thing?" asked Bill.

"Since it's your money, you should decide how to use it. The rest of us are just along for the ride," Eddie said.

"I think we should use it to rent a space someplace large enough for all of us to work on individual projects. Whenever one comes up with something that is marketable we use the proceeds to grow," Bill said.

"Next question; do we want to do that now here in Berkley, or wait until the semester is over?" asked Beth.

"I am not particularly fond of staying here, and I think we could get more bang for the buck so to speak if we found someplace less expensive," Eddie offered.

"I'm with Eddie," Cindy said. "We could go to someplace like Phoenix, Arizona or Albuquerque, New Mexico. We should all go to the place where they supposedly found the Alien bodies back in the nineteen forties. I forget the name of the place, but I know it was in New Mexico."

"I believe the name of the nearest town was Roswell. It's not too far from Albuquerque, and now that we have had our own encounter it would be nice to see if we get any vibes from the place," Bill said.

"Because we want to look at it doesn't mean we have to settle down close by. We can spend a day there on the way to someplace else," Cindy said.

"Albuquerque is a nice place once you learn how to spell it. The summers are hot, but not scorching like Phoenix and the winters are cold but not frigid like Denver. The population is large enough to have all the things we could possibly need and the prices are not out of sight. There's enough open country to get out and hike or camp, and the accessibility is great from almost anywhere else in the states," this from Beth.

"Anybody got anything against Albuquerque?" Bill asked.

No one dissented so he said, "Then let's start looking into housing and office space so when school is out we can go there directly."

They all found their classes boring now. Most of the lecture material they had heard previously somehow came to mind easily and their understanding of the course material was much better. They were all good students to begin with and were now the undisputed top students. They spent a lot of time within their own minds devising and refining ideas for potentially profitable products. Bill already had an idea for a drug to treat AIDS, a virus that attacks the human immune system. If he could pull the right ingredients together he thought he might be onto something.

Eddie meanwhile was still masticating the idea of a thought computer. He could envision how the electric impulses would flow from the brain to the computer but was still working on the other part of the connection. He had done some reading but the books were too shallow to help his understanding of a more complex subject. He would have to bone up on theoretical physics and learn a little more about the flow of electricity and the free flow of neurons, but he felt he was on the right track.

Beth and Cindy were collaborating on an idea for a makeup that was neutral in color but when applied would take on the tone of the wearer's skin. Cindy jokingly called it chameleon powder, but it was an apt name that they all adopted.

By the time school was out and they had arranged for work and living space in Albuquerque they were about decided on the first project for each of them. They furnished the house they had rented with used furniture and yard sale dishes and cooking utensils. The space they used for work was an old refurbished warehouse, much too big for their purposes, but the price was cheap enough that they decided to rent it and build individual work spaces within the cavernous area. Bill would need lab equipment and a computer system devoted to his medicinal research, while the girls would need totally different equipment. Eddie could get most of what he needed for the initial part of his project from Radio Shack or other electronic stores.

The work was invigorating to all of them and they spent twelve to sixteen hours a day in the warehouse. Beth and Cindy spent a good bit of time in the office they had cobbled together researching different topics on the internet and started an advice website. They charged for advice on technical matters and even did some programming, designing websites for a fee. Within three months they were known around the computer world as the best problem solvers in the business. There was not a computer virus in existence that they could not defeat in a matter of minutes. Cindy put together a program which allowed her to remotely access computers to get rid of virus and malware problems. Their income from the computer work grew to the point that it took half their time keeping track of the money coming in through different on-line payment systems.

Beth even wrote a program to streamline the process so that every aspect of the operation was done automatically. She then patented the program and sold it to financial firms from credit card companies to banks and even to government agencies.

Bill and Eddie didn't pay much attention to what the girls were doing until one day Beth said, "Well, how does it feel to be millionaires?"

"What are you talking about?" Eddie asked.

"Our net worth is now over a million dollars and growing every day."

"How did that happen? I thought you girls were doing the computer stuff to keep us afloat until we designed something we could sell?"

"We beat you there. We got a patent on a financial program for online payments that really streamline the transaction process and it has been selling like hot cakes. We have been getting offers everyday to design programs of different types for some high powered companies."

"How are the makeup sales going?"

"Pretty good. We farmed the formula out to a Malaysian company on a percentage basis. They do all the production and sales transactions. We handle the marketing and finances. That has earned almost a quarter million and should double in the next six months."

"And here Bill and I thought we were the brains of this outfit."

"How's your computer idea coming along?"

"I have the brain to computer part of it figured out, but the computer to brain part is elusive. I think it has to do with synchronizing the frequencies between the two components. The brain doesn't receive the return signals on the same frequencies that it transmits and I have to isolate those and the sequence they step through. I understand the concept all right, but the particulars are eluding me. I may need to design a test set and use one of you guys to authenticate the design."

"Well, at least we won't starve while you are working on it. Bill says he has a drug to treat AIDS ready for testing and the clinical trial process. We may have to rent additional space at a different location for his testing. Everything will have to be documented by the FDA and he will probably have to hire doctors to administer the test program."

"I suppose we can afford it, thanks to you girls?"

"It shouldn't be a problem."

The group had been so involved in their own pursuits that none of them had thought about the alien encounter for a long time. Occasionally Bill would wonder again about the aunt who had left the money. Something still bothered him about the whole episode but he couldn't put his finger on the problem. They were all pleased with their accomplishments over the past few months and Bill was starting to look at the stock market for ways to invest their money.

When he found something that looked promising he would research the company on the internet to get a feel for the company's potential. He didn't consider any company unless it produced something technical that would impact the world market. He stayed away from the financial institutions and investment companies in favor of companies that produced tangible products. He considered computer software development companies because they were tied in to computers which were tangible. He found a company that was working on a processor that would almost double the speed of the computer. It was related to what Eddie was doing with the thought computer but didn't involve anything other than the computer and its ability to operate more efficiently.

He approached Eddie with the data he had gathered on the company. "This looks to me like it has promise between now and the time you perfect your thought computer."

Eddie looked the information over and agreed that it held promise. "We will have to find a way to learn how far along they are and what's holding them back, but the price of the stock is so low that it wouldn't take much to push it up. If the processor does what they say it will do the stock could multiply a hundred fold easily. Go ahead and buy some of the stock and I will see about getting a job with the company to help them along with the project."

The company was located in their old stomping grounds in the San Francisco Bay area. Eddie called the company and inquired about the possibility of coming to work for them. He told them he had researched the company and felt that his interests were in line with what they were doing and thought he could be a valuable asset. He was invited to come for an interview.

The following day he flew to San Francisco and rented a car. His interview was with a trio of computer geeks and the head of the company. They talked about computers in general and some of the applications in the past that had failed and some that were huge successes. The head of the company asked Eddie, "What do you think the difference was between the successes and failures?"

"The successes were based upon filling a legitimate need and the failures were based on the engineer's perception of need."

"Could you expand on that a bit?"

"Most engineers have tunnel vision. They get so wrapped up in their own world that they don't see the rest of the world clearly. They focus on engineering accomplishments for the sake of designing something nobody else has done yet, when their focus should be consumer driven. The greatest invention in the world wouldn't mean much if nobody used it. Any successful company concentrates first on the market. What can he build that the public needs and is willing to pay for. He then has to factor the cost to benefit ratio. If it costs more to build the item than the public is willing to pay then sales will be down in the grass. If a person can come up with something that fills a need, that everyone loves, and that he can produce very inexpensively then he has a winner."

"And what can you offer this company that might cause us to want to employ you?"

"I can solve your chip problem for your new processor and get it on the market within three months. I have been working on a similar application for over a year and I know the problem you are having is in the chip circuitry. You hire me on my terms and I will get the product ready for production within sixty days. I don't want any patent royalties or bonuses. A flat fee of half a million dollars is my price."

"And if you can't produce what you say you can?"

"Then you save half a million dollars and probably go bankrupt in the process."

"Surely you can see what this will mean to the computer industry and I can't believe you only want a half million dollars of the huge profits that are sure to follow."

"I am going to buy a lot of your stock before I come to work for you. If the value of the stock increases the way I think it will, that will be my payday."

"What about insider trading laws and such?"

"That's why I am going to buy the stock before I come to work for you. The law doesn't prevent a man from trying to better his investment by working for a company in which he holds stock."

"You're serious. You think you can do this?"

"I am sure enough to put my own money into my conviction."

"When do you want to start work?"

"The day after tomorrow. It is too late today to purchase stock so that will have to wait until tomorrow. Date the contract accordingly and I will start to work in two days time. One other thing, I am going to work in my own lab."

"What's wrong with working here?"

"I don't live here, although I did go to school at UC Berkley for a couple of years."

"What are we supposed to be doing while you are working on this? If you are sure you can solve the problem we will just be wasting the time of a lot of people?"

"I suggest you start to strategize your production and marketing. You might want to contact the major computer manufacturer and negotiate contacts for the processor. Don't try to take them to the cleaners right away. You will hold the patent, and once they see how this affects the industry no one can afford not to use your processor. That's the time to make the money."

"With all you are telling us, you have given this a lot of thought. You could develop the processor under your own patent and make a lot more money than you will be getting out of this, even with the stock."

"The reason I don't want to go that route is that this is going to be a short lived technological advance. Within three years it will be obsolete. I am working on the follow on system."

"In three years this company will make an awful lot of money."

"That it will, and you folks around this table will all be multi millionaires by then."

"I will have our lawyers draw up the contract and date it day after tomorrow. I wish you the best of luck for your sake and ours."

"You won't regret the decision."

Eddie returned to the airport and turned the rental car in. He didn't notice the car that had followed him all the way.

Chapter 3

When Eddie got back to Albuquerque he got together with the others and told them what had happened. "I believe we should load up on the company stock. I can see this coming together faster than I indicated to them. The chip is an easy fix but it will take time to get into production and distribution to the computer makers. We might want to do a software program to back fit this to existing computers. There's a huge market there."

"I could work on that while I wait for the drug to be tested and certified," Bill said.

"And Cindy and I can look at how to market the fix. A download from our web site would be the simplest, but won't the users need the chip installed first?"

"Sure, but you can set up an automatic order form when they access the web site. Tell them that the chip has been ordered and to install the chip per the instructions and access our web site again to get their system upgraded. You will need to set it up with the people making the chip and I will help you with that when I deliver the chip to them."

"How long before you have the chip ready?" Beth asked.

"I could probably have it ready tomorrow but they need time to make plans so there's no real big hurry. I may have to give them some help with the software too. It needs to be streamlined."

"How much of the company stock should we buy?" asked Cindy.

"What's it trading for now?"

"It's under a dollar a share."

"Use your own judgment. I think it will be worth fifty dollars a share inside six months, and probably will continue to climb for the next two years."

"We need to buy the stock in our company name and make sure the transaction takes place before I officially start working for them."

"I will do that this very day," Beth answered.

"Have you guys been looking for any sign of the coming destruction Joe talked about?" asked Bill. "It has not been on my mind much lately and I have an intuition that it should be."

"I have not been consciously thinking about it, but there has been a little tingle in some part of my mind kind of like a sticky note to remind me that there's something I should be doing," Cindy said.

"It's easy to get caught up in our own activities with all the brain power at our disposal. If we still go on the premise that nuclear or biological are the primary modes of destruction then there are no indicators that I can think of. If it is something else I guess it hasn't jumped out at any of us yet," Beth added.

Although they didn't know it, all of them were thinking along the same lines. Why had the alien presence chosen them? Why had he chosen such a remote location to contact them? He said he had been around the earth for ten thousand years, and certainly would have learned a bit more about population centers where he would have a better choice of subjects. The encounter had not taken long at all, in terms of their method of telling time. It had seemed to be a clandestine meeting when all the individual items were factored in. The time, the brevity, the location and the purported reason for the encounter all pointed in that direction. Each of the four had some small element of doubt in their mind telling them that all the cards were not on the table yet.

They were somewhat lulled by the vastness of the potential within each of them. The idea that nothing was beyond their range of capabilities was intoxicating and it was easy to get wrapped up in sheer joy of using their minds to their full capability. If the entity who had contacted them had been around for as long as he said, he would have tried this experiment in the past and there would have to have been some evidence of sudden radical changes in the way of life on earth.

One could point to significant discoveries or inventions to substantiate evidence of previous alien presence, but the scientific and technical advances of humankind would have to be evident over a period of time to validate any previous influence. The truth was that none of them had advanced far down that path of reasoning yet, and it would be some time before they did so.

"We need to make a list of all the precursors we can come up with that would have to happen before any devastating event could take place in either of those categories. That way we could start to look for those things and check each one out for an underlying cause," Bill said.

Cindy said, "I have been doing that on my own. I scan the internet news every day and follow up on anything that even comes close."

"Well, how about we call it a night and go get something to eat?" Bill asked.

They closed down what they were working on and left the lab. As they drove to the restaurant none of them noticed the vehicle that discretely followed them.

For the next few days each worked on his or her own projects. Eddie had done the chip design and the attendant paper work for the patent of the processor. Though he had not seen the design of the system the chip was to work with, he was sure the circuitry would run it. He waited a few days to call the company and tell them the chip was ready. He would deliver it in person and install it in the prototype system they had developed.

Bill in the meantime was away from the lab a lot, setting up his procedures for the testing of the AIDS drug. He would need to find test subjects from the population of

HIV infected people. The test plan he had put together had already been reviewed by several physicians familiar with the steps necessary to get FDA approval and had people searching for a good location to administer and follow-up on the test subjects. If he could get the FDA to fast track the tests the drug should be ready in six months.

Eddie was preparing for a trip back to the Bay area to deliver the chip and would be gone for a couple of days. Beth and Cindy would be alone in the lab for those two days and were working on the marketing plan for the processor system Eddie was involved with. It was nearing lunch time on the day Eddie had left for San Francisco and Beth said to Cindy, "I'm going to go get us some lunch. What do you feel like eating today?"

"I don't know, maybe a salad and some yogurt."

""I will be back shortly."

Beth grabbed her purse and left the lab, locking the door behind her. Because they could not view the entrance directly from where they worked, they usually kept the door locked. In addition there was a camera mounted to allow constant surveillance of the entrance. She went to her car and was about to get in when she felt an arm go around her throat and a hand grab her arm to twist it behind her back.

She reacted instinctively and spun into the direction the assailant was pulling her. With her right hand, which was free, she latched onto the wrist of the arm encircling her neck, bent her knees and yanked the assailant over her head. There was a cry of pain as he struck the ground and she was about to follow up with a disabling blow when she felt the barrel of a pistol pressed against her spine.

"If you make one move, you're dead. Put your arms straight out."

Beth complied with the order.

The other man was getting to his feet and went around behind Beth.

"Now very slowly, bring your hands behind your back, one at a time."

Again Beth complied. She had no idea what was going on. Her hands were cuffed behind her with plastic handcuffs. She was then taken to a car and pushed into the back seat. One of the men got in behind her and the other got behind the wheel. She was still trying to figure out what was happening when the man in the back seat jabbed a syringe into her upper leg. Shortly thereafter she lost consciousness.

She was taken to a warehouse similar to the one they had rented for their laboratory. The drive-in door was raised and the car drove inside. The two men got her out of the car and carried her to a room within that was outfitted with a makeshift bed with a lot of equipment surrounding it. They placed her on the bed and a third man began attaching leads to her head in a fashion that indicated that he knew what he was doing. Once all the leads were attached he went to the control system and punched a few keys on the keyboard. An oscilloscope showed waves much like those on an EKG scope. The operator studied the monitor for a minute or more then input different commands into the machine. He didn't print anything, just looked at the different readouts. After about fifteen minutes he said to the other two, "Take her back and put her in her car."

They complied with the order and placed Beth behind the wheel of her car with her purse, and left the keys in the ignition.

When Beth regained consciousness she was extremely disoriented. She finally realized where she was and tried to remember what had happened. She remembered the two men and them loading her into a car but she remembered nothing thereafter. She rubbed her leg and felt the area where they had administered the drug. This caused her to remember being stabbed with a needle. Everything beyond that was a blank. She had no idea where they had taken her and looked at her watch to gain some indication of how long they had her. Less than an hour had passed, so they couldn't have taken her very far away.

The more perplexing question was why they had attacked her. It obviously wasn't to rob her since she still had her purse. They had not physically harmed her, except for the injection, which she thought might have been to keep her quiet while they transported her. She had no doubt that they had moved her because the area was too public for an hour to have passed without drawing attention. Beth was hungrier than ever now, and though she was still shaken from the experience, she went to get the food she had originally started out for. When she returned to the lab Cindy said, "I was beginning to wonder what happened to you. You sure took long enough."

Beth related as much as she could remember about what had happened to her. "I don't have any idea what that was all about? The fact that they didn't harm me, even after I threw the one to the ground makes me wonder what they did to me while I was unconscious."

"We have to tell Bill and Eddie about this right away. It might have something to do with all of us, not just you."

"I was thinking along the same line. The fact that they didn't harm me makes me think they might have gotten what they wanted, whatever that was. It would mean that they probably hooked me up to some sort of machine to learn whatever they wanted to know."

"And taking that a step further, it means that someone knows about what happened to us or at least suspects something along those lines."

"Could that be the evil twin surfacing? I thought Joe said they couldn't interfere directly."

"Maybe the other entity operates differently. Maybe whoever he has working for him hired someone to do the dirty work."

"I suppose that could be true. The episode makes me want to go back over all we have done since the encounter item by item. Something just doesn't add up. We all seemed to accept what Joe told us at face value. Maybe we were disarmed because he showed no open hostility. He obviously could have done whatever he wanted with or to us during the experience," said Cindy.

"Well, we are going to be alone for a couple of days, and I don't think we should do anything alone. What I mean is, we shouldn't be out by ourselves until we can figure this thing out," Beth added.

"I agree. I also think it might be a good idea to shelve what we're working on and take a logical approach to solving our dilemma. Maybe if we can get all the data available down on paper we can make more sense of it. I hate the thought of being a pawn to some extraterrestrial's whims."

"Let's start right now," Cindy said. "You make a list of all the things you think could possibly be connected not supported by fact, and I will do the same thing for the factual part."

"Okay, but there's no way I can think of to insure the list is all inclusive."

"That doesn't matter. Just try to get us something to work from."

Both women went to their computers and sat down to the tasks. With their enhanced mental ability and near photographic memories they had a lot of material to sort through, and it took the entire afternoon.

Beth's list included all the things they had done to devise ways to make money. Eddie's idea for the think computer and his work on the chip was included on both lists, as was Bill's work on the AIDS vaccine. The thinking was that while neither of them had come to fruition yet, both were sure bets. The big question mark was the inheritance which appeared to be factual, but could have been engineered in some way that they did not know about. Something still didn't add up about the timing and circumstances of that event.

Cindy meanwhile, had listed the encounter with the alien and the circumstances. The fact that they were two couples she listed with a question mark. The events that took place during the encounter were listed chronologically with special emphasis on the alien's reaction to Eddie's questioning about motive.

They finally sat down with the two long lists and attempted to graph or collate the items each had listed. Strangely, neither of them seemed overly concerned about the attack on Beth earlier in the day. They accepted it as evidence that someone or something simply wanted to learn more about them. They accepted the fact that they were sailing in uncharted waters and the way to deal with the problem was cerebral and that they would eventually find the answer.

If someone had asked any of the four a year ago if they believed in aliens, they would all have said no, or at least a qualified no. Since the experience, all had obviously been converted to believers. Their enhanced abilities alone would have done the trick, but the things they had been told about the earth's destruction, or actually, lack of things seemed much more plausible to them. What didn't seem logical was that some rival entity would build things up simply to destroy them.

The root of evil always seemed to be money, or more accurately, the things money could buy. Lumped into that category was almost every human trait. Power was always the driving force, and he who had money had power. The power could be used to rule kingdoms as in days of old, or in modern times to bend the wills of people a given way. If the objective was destruction of the world it made no sense to even bring money into the equation. Simply build enough weapons of mass destruction, whether nuclear, chemical, or biological and get it over with. Entities such as they had encountered surely had the wherewithal to accomplish that. They probably even had some sort of death ray or unheard of weapon that could accomplish that without even having to come close to the Earth.

Finally Beth said, "It's hard to see any pattern in all this, but the one fact that stands out, is that we are being observed, or followed, and I would be willing to bet that it has been going on since our encounter with Joe."

"I have to agree with that; but why, and who, and how many? Have they been following all of us, or just keeping an eye on our lab? I want to do some more digging into Bill's bequeath, not through the lawyer, but through state records. I want to see a death certificate on his aunt and a list of her holdings. The will would have to be probated through the state, and I will give you good odds that when we look we will not find a death certificate or a will. Someone knew we needed the money to get started and found a way to provide it that we wouldn't look at too closely," Cindy said.

"If that's true, and I'm beginning to believe that it is, then who orchestrated it. Joe? He seems the most logical, but if he told us the truth about not being able to interfere, then it had to be his adversary, and how would he know about us? I think we should contact Bill and Eddie and at least warn them to be on the alert to whatever we're up against."

"I agree with that one hundred percent. You know, the way we could get some answers is to go back and talk to Joe again, or commune with him, or whatever it is we do."

"That's something to think about, but we really need to talk to the guys and see if we can make some sense of all this first. I'm going to call Bill, and you call Eddie," said Cindy.

When Cindy got Bill on his cell phone she told him that he should come back to the lab as soon as possible. She didn't tell him anything about Beth's experience, but she did ask him to see if he could determine if he was being followed by anyone.

Beth was more open with Eddie. "Someone assaulted me this afternoon in our parking lot and drugged me. They didn't harm me, nor did they rob me. When I woke up I was in my own car in our parking lot after a little over an hour had passed. Cindy and I have been going over everything we can recall about the last few months and things don't add up too well. When we are all together again we need to try to make some sense of it. For the time being I wanted to tell you that I think we are all being followed."

"You think someone might be following all of us?"

"I don't think they would just single me out. They jabbed me with a hypodermic needle and I don't know what they did or where they took me, but we couldn't have been in the parking lot all that time. Someone would have noticed. So the conclusion I draw from this is that someone knows about our encounter and is trying to gather some information, but I have no idea what it might be."

"I will see if I can find out if I am being followed. I don't really need to spend much time out here. If I can wrap it up this evening I will fly back tonight. Get in touch with Bill and see if we can all get together first thing in the morning."

"Cindy just talked to him and he should be back at our lab later today."

"I will talk to you later today and let you know when I will be getting back."

Cindy and Beth stayed in the lab until Bill showed up just after six o'clock.

"Well you girls are certainly onto something. I think I was followed, but I can't be absolutely sure. The car I pegged as the follower didn't make the last turn with me, and I waited in the parking lot for a few minutes and didn't see any evidence of it in the area. He stayed within sight distance of me all the way back until the last turn though, and may have broken off when he determined my destination. If that's the case, it means that they have someone watching our lab, probably from a fixed location."

"Beth talked to Eddie and alerted him. You know, if they are following Eddie in San Francisco, it means that we are dealing with someone with a lot of bucks. It can't be cheap to field enough manpower to try to keep tabs on all four of us all the time," Cindy said.

"At the very least it lends credence to what Joe told us. I'm not saying that it validates his claim of coming world destruction, but it does point to the existence of the evil twin," replied Bill.

"As soon as Eddie gets back we have to sit down and figure out what all this means."

"We might as well just cool it until then. In the meantime let's all give it some thought and try to find a direction to take for resolution of the problem."

Chapter 4

Meanwhile in a high rise in New York City a heavy set man in his middle sixties sat in a high backed executive office chair turned away from the desk and facing the view of the city through tinted windows. He had a phone to his ear, listening to the caller. He said into the phone, "And you didn't find out anything at all unusual about her? Not even enough to determine if she was extra smart?"

The man on the other end of the conversation said, "I hooked her up to the machine like you told me, but only normal brain waves showed up. There was nothing to differentiate between hers and the average person's. The only thing that might indicate that we're on the right track was the way she reacted when Carl and Eric were picking her up. Carl, who is pretty strong, and highly experienced, tried to make the grab and she threw him completely over her head and onto his back. If Eric hadn't put a gun in her back she probably could have taken both of them, and that's out of character for a petite lady who has no training or experience."

"Well, stay on them. They're the only likely candidates that have come along in the past ten years, and I can't afford to let them slip through the cracks if they're the ones."

"Is there some other way we can tell for sure, other than the method I used on the girl?"

"The only other indicator will be if they start to accumulate a lot of wealth over a short period of time. They will be capable of doing that in many ways, but my guess is that it will be in the technological area. Extraordinary intelligence is hard to harness. It makes you itch to try out new ideas and theories. If they're the ones, they will not be content to sit back and take the easy road to riches; they will come out with new inventions and improve on a lot of the current stuff. I need more time to make my own preparations, so keep a watch on them and if you need more manpower or money, let me know."

After he finished the call the man, whose name was Clarence Woodman, turned back to his desk and sat in thought for a long time. He was mentally reviewing the last forty years of his life, wondering if the latest subjects could be the ones he had been searching for. He was wealthy beyond his wildest dreams and had everything a man could want. His fortune put him in the top one percent in the entire world, where he had been since his encounter with the alien. He was only in his twenties at the time and was probably headed for a life of crime. He was not well educated, hated physical labor, and loved the finer things in life.

He had a run in with some enforcers from a bookie joint in his old Philadelphia stomping grounds and decided to move to New York and try his luck. He had been hitchhiking and was on a lonely stretch of road; it was getting dark and he was tired. He spotted a barn near the road and decided to go there and sack out for a while. He was moving through a grove of trees when he suddenly had the feeling that he was not alone. He was armed, so he pulled out a pistol and slowly turned around in a complete circle, looking for anyone or anything. There was nothing there that he could see, so he returned the gun to his belt. He still had an uneasy feeling and if the truth be known, was quite afraid. There was nothing that he could see, yet he intuitively knew that he was not alone.

As he stood there he seemed to be floating away from his own body. He became aware that something was happening, but it was something that he had never experienced, almost dreamlike, yet his senses seemed to be on high alert. A voice, at least he thought it was a voice, said to him, "There's no reason to be frightened. I mean you no harm."

As he heard the voice he became aware of a misty presence. It seemed to have substance, but was opaque.

He couldn't even discern the true shape. "Who are you? What are you?"

"I am what you earthlings call an extraterrestrial. I am not from your planet, not even from your solar system in fact. I exist in what you would call spirit form, although I appear this way because it is the form that is less likely to elicit a negative response from you."

"You're real? This isn't just a figment of my imagination?"

"I am quite real, though reality to you and I do not necessarily mean the same thing. I come from a solar system far away, but I have been in this vicinity for thousands of your years."

"What do you want with me?"

"I am sort of playing a game with one of my kind. I am trying to devise a way to accumulate all the wealth on your planet, and he is trying to thwart my efforts. Would you be willing to assist me if I could make you the richest man on earth?"

"The richest man on earth; you've got to be kidding? I would do almost anything for that."

"Well, we might be able to do business. I will provide the tools for you to do this, but you in turn must follow my orders. My adversary will be trying to do the same thing. I will be at a disadvantage because I can only have one of your kind working for me while he can have many. You will have to do all the things I tell you, and look for the signs I will explain to you. Would you be willing to do this?"

"Of course. You did say the richest man on earth?"

"Maybe I exaggerated a bit, but you will be in the top ten."

"That would be more than I will ever need. How are you going to do that?"

"I will provide the tools for you to do it yourself. No one will ever know of my involvement. Do you agree that you would like to participate?"

"Hell yeah. What do I have to do?"

"I am going to place an object over your head. You will not feel any discomfort, but the object will alter the makeup of your brain. You will become much smarter by a magnitude of ten times, maybe more. You will have the intelligence to excel at whatever you choose to do in your world." With that he placed the helmet like object over his head.

"I didn't feel anything."

"You weren't supposed to. Now, I want you to go on to New York. You will know what to do when you get there. I will not be able to communicate with you except from this specific spot. Once each of your earth years I want you to return to this very spot to commune with me. If circumstances require it you may return sooner. Once you gain your wealth, I want you to be on the lookout for people who possess the same abilities you have been given. It will likely be more than a single individual. I do not know when they will appear. It may not even be in your lifetime, which will mean that you will live a life of luxury without encumbrances."

"That's all you want me to do, look for these people?"

"And inform me when they appear. I will tell you what to do after that. The reason for your wealth is so that you can employ the people and means to do the chore. If you need an army, hire an army. If you need a country, buy the country. Spare no expense in carrying out this task."

"This sounds too good to be true."

"However, it is quite true."

"When do I start?"

"You already have. Now you need to start employing the tools that I have given you. I will see you in one year, right here."

And for the next forty years they had met. Now his people had identified four young people who exhibited the characteristics that the alien had told him to look for. He had people looking for over forty years, and this was the first instance that gave any credence to what the alien had told him. He had come to doubt the truth of what the alien had told him, at least in terms of it being a game between himself and another alien, and couldn't guess at the motive of his benefactor. Everything he had told him about the wealth part of the deal had come to pass and he had lived a life for the past forty years that he could never have dreamed of prior to the meeting on that fateful day. His was a life of leisure which gave him much time to consider his situation from all angles. First of all, he wondered what he would be told to do if he should somehow manage to find people like himself. They would have the brainpower he did, and might not be so easy to deal with. He wondered what would happen to him if he just forgot all about his encounter with the alien. Since he was able to bestow the powers he had on him, he surely must be capable of doing him bodily, or mental harm if he so chose.

There was an element of fear about the end result, no matter what he chose to do. If he did the bidding of the alien, he would probably be asked to get rid of the newcomers, and that might lead to trouble with the law. He could hire people to do the dirty work if it came to that, but he had been away from that shoddy world for such a long time that he no longer had a stomach for it. He would have to wait and see how the current situation played out.

Chapter 5

Eddie had returned from San Francisco late the previous night and they all sat down together after breakfast to discuss the situation. After Beth had called to warn him about possibly being followed, Eddie had paid more attention and had detected the car following him in San Francisco. He had also noticed that he was followed from the airport when he got back to Albuquerque.

"I was followed to the airport in San Francisco, and followed from the airport after I landed. Whoever is behind this must have a lot of money, because he is certainly employing a lot of people. If what Joe told us is true and they can't interfere directly, then whoever is behind this has to have been involved long enough to accumulate enough wealth for his purposes," Eddie said.

"When Joe said they couldn't interfere directly did he mean physically and mentally? If we approach the problem based strictly on facts, and on what we know of the laws of physics, I think the conclusion would be that to interfere physically, some scientific laws unknown to us would have to be employed. And the mental part of it would seem to be improbable because he had to do something with whatever contraption he used on us when we met. If he could influence our mental process remotely, then there would have been no need for the meeting. On that basis it would seem that the other side can only do what Joe did with us, but he could give them directions about what they are to do," Bill said.

Cindy, who had been turning the matter over in her mind since Beth's incident the day before, had come up with one important additional question. "If what you say is true, then how did whoever is behind this latch onto us so quickly? We haven't done anything to call attention to ourselves. And if someone for the other side has been in place for a long time and Joe didn't know about it, then it stands to reason that they could not know about us."

Eddie tried to express the situation in a way that they would better understand it. "If Joe's adversary has had someone in place for a long time, then he may be watching for signs that would indicate someone like us has come along. While he may not be able to interfere directly, he could still give him indicators to look for that might point to his opponents being on the scene. I still don't understand how all this is supposed to tie in with the destruction of the earth."

"Maybe it doesn't," said Bill. "All this could be a natural result of the rivals putting all the pieces into play."

"If that's the case we have to concentrate on staying alive for the main event, if it comes during our lifetime," Beth added.

Eddie said, "You know, even considering everything that has happened so far, I still have a hang up on the destruction of the earth thing. Based on what we all know now, and even with the enhanced brainpower, no scenario I can think of supports total world destruction as a possibility. Even if there was a nuclear Armageddon there would be isolated places on earth that should survive. There are so many countries with nuclear weapons now that it is just not conceivable that all of them would use them at once. If one of the major powers, with a vast nuclear warehouse, unleashed an attack, some of the countries would be wiped out before they could respond. That would reduce the number of bombs actually detonating, and you have to figure they would be detonated over populated areas. Other spots that are desolate, like Antarctic, jungles in Africa and South America, and island nations should survive, so total destruction would be very hard to achieve. The biologic aspect would spread under the same constraints. The more I think about it, the more doubt I have that Joe was truthful about that part of it."

Cindy said, "Maybe this is a bit reckless, but what do you guys think about going back for another session with

Joe? I don't think we are going to be able to figure this out on our own. If we just go back and say 'hey Joe, we think you fed us a line about the earth's destruction' he may come clean with us."

Bill responded, "That may be the only way to get answers to our questions. We have puzzled over this long enough, and we still don't know what we are up against."

Eddie said, "Now that we know we are being followed we can take steps to negate that, at least to some degree. I don't subscribe to the theory that simply because they didn't harm Beth when they had a chance that they only mean to observe us. It could be that they, whoever they are, want to make sure before they do anything really drastic. I believe we should take some steps to protect ourselves as much as possible. For starters, we should get some firearms training and apply for gun permits. I don't believe in this state that you even have to have a reason, just not have a criminal record. We can do that much and make sure we are not alone in a situation where they might try something."

Cindy said, "That will help I suppose, but I think it is more a psychological advantage than anything tangible. If they plan to get rid of us I wouldn't expect them to mount a frontal attack. They would be more likely to try something like burning the house down with us in it, or the warehouse."

Bill asked, "Eddie, do you have to go back to San Francisco any more for the chip design, or did you test it while you were there?"

"I put it in the prototype and it worked fine. There are some software issues that need to be addressed, but I can do that from here and either e-mail them, or send them overnight mail. I don't have to be physically present for anything more to do with the project."

"I have everything lined up for the drug trials. I hired four doctors to run the test program. They are all young, except for the test director. I figured it would be better to have someone with a proven track record for previous tests of this nature, and someone who knows the pitfalls and how to deal with FDA inspectors. The test will take six weeks, but the follow-up tests have to confirm the success of the drug for six months thereafter. I don't have to be involved in any of that, except to pay the salaries of the people involved. We will also be paying the test participants. I have been thinking about the possibility of finding a drug company now, so that when the drug is ready we can go into mass production right away."

"What kind of percentage would you have to pay a drug company to produce it?" Beth asked.

"I haven't checked into that very far. You might want to research it for us, and determine if it would be better to sell the patent to the drug company outright. You could also look at that from the standpoint of a percentage of sales revenues. Factor in what we spent for development and testing and get that money up front. I don't see us getting tied down to the manufacturing and distribution of the drug."

"I will do that within the next couple of days," Cindy said.

"I am still really hung up about the people watching us. If we assume that someone has been set up by the other alien, and long before we were, then he will be extremely rich and that gives us a starting point to look for him," Beth said.

"Or her," Eddie added. "We can't just assume that it is a man. If whoever it is was approached in the same manner as we were, then they will have the same advantages we now have. And if they have been in place for any length of time they will have been told what to look for. That still doesn't settle the issue of what this whole thing is about though."

"I still think the only way we are going to get to the bottom of the puzzle is to go back to Joe and see if we can learn anything more. Between the four of us you would think we could figure out anything that is based on logic, and none of it makes any sense," Cindy reiterated.

"How do you all feel about doing that?" asked Eddie.

All shook their head in the affirmative. They would have to plan it in such a way as to throw off their watchers. They kicked the subject around until they came up with a plan that they thought had the best chance of success. But before they did any of that, they were going to buy personal weapons for each of them and get carry permits.

The next several days they kept a very sharp eye out for the followers every time one of them went someplace. They wanted to see how much manpower the other side had dedicated to them. They left the office individually a few minutes apart to see if they were all picked up immediately or if some were followed and some not.

On the first run Eddie, who left first, was followed, as was Beth, who left next. Cindy then left and was not followed, but Bill, who left last, was also followed. That meant at least three people were with the group watching them.

When all of them were back at the warehouse Eddie told them, "I have an idea. Let's see how good they are. I would like to find out where they are watching us from. In order to do that one of us needs to lose them after we leave here and someone else will pick them up and follow them back to their base. I imagine they coordinate by cell phone to keep track of us."

"We need some form of transportation they won't recognize in order for this to work. Why don't two of us leave together and the passenger can duck out of the car someplace along the way and rent another car, or motorcycle and call when they are ready. The other can lead them by wherever we want to pick them up and follow them back toward the warehouse. It seems that they always leave us alone after we make the last turn. That should allow whoever is prepared to see where they go after that last turn," Beth said.

Cindy said, "Wherever they are located, it has to be with a sight line to our building, otherwise they wouldn't know when we were leaving until we actually turned the corner where they pick us up. Bill and I will try to do the counter surveillance thing and you guys see if you can pinpoint any location that has a direct line of sight to where we exit our parking lot."

"Something else you can do while you are out; buy a pair of strong binoculars," Beth said.

"While you guys are gone, I will start looking for the names of the wealthiest people in the world. I will see if I can get a short biography of each of the filthy rich, and we can start to look for the most likely candidates," Eddie said.

They put the plan into motion right away. Cindy and Bill left the building and Eddie went out the back door of the warehouse opposite from where Bill and Cindy would exit and went to the corner where he could stick his head out and get a look at the entrance to their building. As Cindy drove away, he watched the street to see if the car picked them up at the corner where he thought they would be likely to come from. Sure enough, within a few seconds of their passage, a car turned out behind them.

Eddie called Bill on his cell phone and told him he had seen where the car came from and what kind it was. Bill replied that they had detected him as well. Cindy was going to a shopping mall and drop him off. He would then find some mode of transportation and call her when he was ready to tell her where he would pick up the following car.

After the phone call, Eddie looked closely at the area that would be down the street the car had come from. Most of the structures were warehouses similar to the one they used. He did not see any windows that looked to be likely places for anyone to watch from. It could be that they had a camera set up to watch and the monitor was inside the building. He would need the binoculars to search the area closely.

Eddie called Bill again. "When you start back give a call with your ETA and I will put on my running shoes and see what I can tell that way. I may be able to hide and observe which building they go into."

"I will do that. I suppose you still want the binoculars?"

"Yes, they will come in handy at some time or other."

He went back inside and told Beth what he had in mind, and that the street they thought the watchers were using had been confirmed. "I didn't see any windows that gave a good sight line. It is my guess that they are using a camera mounted on the roof or side of the building. I should be able to tell that with the binoculars, if that's the way they are doing it."

"What are we going to do if we locate them?" Beth asked.

"That's a bridge we will have to cross later. The first step is to locate them and determine how many there are. The next logical thing would be to see how they are set up. I expect that you were taken to that location when you were accosted. You said that you were only out for a little less than an hour, so they couldn't have gone far. I would further expect that what they did was try to see if they could detect anything different about the way your body or brain functioned. You know, that's something I hadn't considered before. None of us have been to a doctor since the encounter and I wonder if there are any differences that will show up on the normal tests done by medical people?"

"We can check a lot of those things on our own; like respiration, blood pressure, body temperature, and reflexes. The things like blood chemistry, brain wave patterns and molecular tests would have to be performed at a hospital, or at least by someone who had the proper equipment," Beth said.

"I would bet a dollar to a doughnut that is what they did to you. Apparently someone is trying to find out if there is anything different about us, and you were the guinea pig."

"What you are suggesting is that Joe's evil twin has had someone in place a long time and he has been on the lookout for God knows how long for someone like us to come along. That would validate Joe's statement about an opponent, but I still have trouble with the rest of the story."

Eddie continued along the path his thoughts were taking. "I don't believe any of the stuff he told us about the destruction of the earth. Some calamitous event might happen, but I don't believe it will be caused by anyone on the earth. The earth could be struck by a large meteor hard enough to cause its orbit to change, which could result in drastic and immediate climate changes. This would probably affect the ozone layer and take away the planets ability to sustain life. There are other scenarios that fall into the 'what if' realm, like the pressure inside the core of the earth exploding. The bottom line to me is that it is almost impossible for man to do anything that would totally destroy the earth."

Beth had been pecking away at the computer during their conversation, and a laser printer had been printing almost constantly. She now gathered up the stack of paper from the printer and started to sort the documents she had printed. "I have the top twenty five richest people in the world in this stack. We will start with them and expand downward until we narrow the list down. I have a feeling that the person we are looking for will be in the top one hundred if he has been endowed the same as us."

"We should be able to learn a lot just looking at how those people made their money," Eddie said.

His phone rang and he looked at the caller ID. It was Bill telling him that they would be returning shortly if he wanted to don his running shoes as he planned. He briefly told Beth what he had in mind and changed shoes. Beth locked the door after he left and he started jogging away from the street where the car had come from that followed Bill and Cindy. He planned to circle the block and time his approach to the time Cindy would be getting back to the warehouse.

He called Bill again and asked him to give him a few minutes warning. He got the return phone call within ten minutes. He had loitered on the opposite end of the block so that he could time his approach properly. As he slowly jogged down the block he saw Cindy pass and within a few seconds a car turned down the street he was on. He continued jogging and watched where the car went. An overhead door to one of the warehouses opened and the car drove inside. Eddie continued to the location and took note of the particulars of the building. He counted the number of warehouses from the corner so that he would be able to pinpoint it from their own location when he got back, then continued around the block and finally back to their place. Bill pulled in with the rental car about the time Eddie got back.

Once they were all inside Eddie told them what he had discovered. "The car turned into one of the warehouses on the next street down. I counted the buildings, so I know which one they are using. There are no windows that would allow them to see us, so my guess is a camera either on the roof or mounted to the side of the building. Let's take the binoculars and see if we can determine anything."

Eddie took the binoculars Bill had purchased and went outside to the same corner of the building he had been at before. He assumed that the camera would be focused on the entrance to their offices, but was cautious about showing much of his body anyway. He slowly scanned the side of the building he had identified and saw nothing. He next panned the roofline with similar results. He was beginning to think he might have been mistaken about how they were doing the surveillance when he noted a mechanical building further back from the edge of the roof. He focused the binoculars and found the camera mounted along the side of the enclosure.

When he got back inside, he simply said, "Bingo."

"Where do they have it mounted?" Cindy asked.

"There's a small building on the roof, probably a mechanical structure for the air conditioner. It is on the upper part of that. We now have to decide what we are going to do about it," he said.

"Well, we know they have guns, because the guy who abducted me had one in my back," Beth said. "Before we do anything, we need to be armed."

"That's a given," Bill said. "I believe we should just let it be for a couple of days while we get weapons and permits. We can discuss the possibilities as we go over the material Beth had compiled."

"Before we do that let me download applications for gun permits. We can fill them out ahead of time and when we go to apply we won't have to sit around half the day to get them. We also have to buy guns. Anybody have any preferences?"

"Something that won't knock me off my feet if I have to use it," Cindy said.

"Something that will put whatever you shoot down with the first shot: maybe a nine millimeter. There's not much kick, and the slug is pretty powerful," Eddie said.

"Let's go over these profiles for awhile and see if anything jumps out at us. We can then go get the guns and stop by for the permits tomorrow. Then we need to go out in the desert to learn how to use them," Beth said.

"I sent the final software changes to San Francisco. They should now be ready to start the manufacturing process for the chip. The entire package will be ready next week. They already have orders from some of the major players. I expect they will send me a check next week."

As they went over the material Beth had printed they were still not sure exactly what they were looking for. "I think we need to establish some criteria for identifying suspects. According to Joe, only one man would have our abilities, and based on the fact that he already has a hefty bankroll to employ the number of people he is using here, he would have needed quite some time to build the bankroll. I believe we can forget about the Johnny come lately types of the dotcom era. Look for someone already established before the computer revolution," Eddie said.

"If he is as wealthy as we think, then he will have built his fortune through stocks and property investments. He may have some inventions patented, so we can look at the patent holders after we narrow the list down a bit. He will also probably have a shady past, and it would surprise me if he is married. We want to look as far back as we can before these people became rich," Cindy said.

They studied the twenty five files Beth had downloaded and pulled the ones they thought met the criteria they were looking for. Of the twenty five, eight were rich as a result of the dot.com boom and were immediately discarded. Several more had inherited the money from family businesses and still others they thought were outside the criteria for one reason or another. When they had all looked at the files, they had agreed on three of the potentials as worthy of further research. Clarence Woodman was on their list.

They finally called it a day and went to dinner on the way home. The same car followed them home.

Chapter 6

Eddie went shopping for the guns and ammunition. He took the permit applications they had downloaded and filled out with him and purchased the guns in each of their names. It was not exactly legal to do it that way, but the store owner didn't want to lose that large a sale and accepted the personal information on the application to submit for record checks required for the sale of the weapons. Eddie paid with a credit card, which eased the owner's mind somewhat. A person who planned something illegal was not likely to buy the guns with a credit card that could be traced back to him.

He went back to the house and picked up the other three and they headed to the desert outside the city. They had explored enough since they had been there that they knew a good location for target practice, and the drive would make the following car standout if he followed them all the way. Eddie kept a close eye on the rearview mirror as he drove.

The location was on the south side of the city and as the traffic thinned, the car following dropped farther back. When Eddie took the dirt road into the desert he chose not to follow, which was fine with all of them. They would still keep an eye out to see if he changed his mind.

All four of them had some fundamental knowledge of guns and Eddie, being the most familiar, showed everyone how to load the clips and insert them. He went over the details of the safety and that the trigger must be pulled for each shot fired. In other words just holding the trigger down would not cause the weapon to keep firing. This was a semi-automatic. As long as there was ammunition left in the clip, the gun would fire each time the trigger was pulled. He put a clip in one of the pistols and demonstrated. They had brought along some tin cans and threw them out in front of them to have targets.

Instead of all firing at once, they took turns. Apparently the enhanced physical ability extended to the hand-eye coordination as they were all fairly accurate with their shots. The main thing was to get a feel for how the weapons felt when being fired and how to align the sight picture. They burned through three boxes of ammunition before they were finished.

The next stop was the court house to apply for permits for the guns. Technically they could just strap them around their waists in a holster and would be completely legal, but if one wanted to conceal the weapon, then a permit was required. The process was not as lengthy as they anticipated, and they were at the office long before noon.

Eddie had a call from the company he had been working for and he returned the call. They told him that everything was set with the system and they had enough advanced sales to pay his fee. They could handle it electronically and the money would be in his account that afternoon.

"Okay," Eddie said, "now let us see if we can get more information on the people we looked at last night. Beth, you try for the deep background on each of them. Go back to their birth if possible. Cindy, you look at their financial records insofar as possible. Bill, you look for dirt on any of them and I will look at the good side, philanthropy, support to the arts and education, that sort of thing."

They spent the entire afternoon on those tasks. At five o'clock Beth said, "Let's put this together and see if we are any closer to figuring this out."

Beth had been able to get ages, dates and places of birth and at least some background on all of them. Woodman was the enigma. There was nothing worthwhile on him until he suddenly burst on the financial scene in his early twenties. He had been born in Philadelphia, was the product of a broken home, and was a high school dropout. He had no record of employment that she could find in the Philadelphia area. The other two were not sterling characters, but the things she found out showed their actions to be outside the parameters they would expect someone like them to adhere to.

Both had multiple divorces and off spring that had made some bad choices. Woodman, on the other hand, had never married and was now in his sixties.

The information Cindy had compiled on each of the three showed them to be diverse as well. Woodman was the only one who had never suffered a financial setback of any type. His climb up the financial ladder was always upward. He had never suffered any sort of financial reversal in the forty plus years that they had been able to track him.

Cindy had hacked into the IRS computers and gotten the tax returns for all of the three. She printed the ones she thought relevant and saved the rest to look at later if need be.

Bill's research had turned up dirt on two of the three, Woodman being the exception. The other two had been involved in law suits from ex-wives and from other companies, though neither of them had ever been convicted of anything. They had settled cases out of court involving companies they owned, but nothing spectacular. Woodman had never been sued, nor had he been involved in anything the press got their teeth into. That was an anomaly in and of itself for one as rich as he.

While the two who had questionable characters gave generously to charity organizations, Woodman had never given anything to anyone, unless he did it anonymously.

"When you put it all together, Woodman has to be the guy we're looking for. He has no record of doing anything until he suddenly shows up in New York and starts to make money in the financial market. That would mean that he is about forty years ahead of us, and apparently has been waiting for our appearance at the direction of the other alien," Bill said.

"Okay, if we are agreed that Woodman is the culprit, what are we to do about it?" Beth asked.

"There's not much we can do at this point. He is untouchable as far as we are concerned. We can break up his gang that is watching us, but to what avail? That will simply confirm to him that he has found the people he has been looking for. I believe we are going to have to have another session with Joe," Eddie said.

"I don't know why, but I am dreading this confrontation," Cindy said.

"I believe what you are dreading is the unknown. None of us know enough about what we have gone through to be comfortable with any of the conclusions we have reached. There are enough anomalies and unknowns to cast things in a different light, depending on whether or not we believe what he told us. We need to get it all out front and decide if he is being truthful with us or if we are just his pawns in a game of world chess," Beth said.

"We know that part of what he told us is true. There is in fact an opponent, and he has been hard at work for forty years. I still can't see a reason for all this. Every time I start to think Joe is just full of it, something happens to make him seem more on the level. I think the only way to settle the issue is go camping again," Eddie said.

"How are we going to get rid of the ones watching to get away for the encounter?" Cindy asked.

"We first need to decide how we are going to get there. If we drive, it will take a couple of days. If we fly into someplace like Fresno or Tahoe, we are still going to have a long drive. If we go in the truck I can lose them easy enough with the four wheel drive, but they might have other methods to keep track of us. If we lose them here, then they will have a couple of days to find us before we can get to Yosemite again. If we fly into someplace we will have to rent a truck and then buy everything we need for the trip. I don't know if it is critical that we keep the location secret or not. Joe didn't mention anything about that but I think we should err on the side of caution," Eddie said.

"We need some down time anyway. Why don't we drive part of the way, like we are going to Lake Tahoe, and lose the tail when we get close? We can get to the site in a day from Tahoe," Bill added.

"The only problem with that is that there are not that many ways in and out of there. All they will have to do is set up on three or four main roads and wait for us to show up," Beth said.

"Then let's drive to California and lose them someplace like Fresno. It will appear that we are going to the San Francisco area if we do that. It will also be a shorter distance we have to drive after we ditch them," Cindy said.

"I think that's the best plan. We won't have to be overly worried about them until we get close," Eddie said.

They decided that the first thing the next day they would pack up and head for California. No matter how they chose to lose the ones following them it would not be hard to do. All the preparations were made that night and they set out early the next morning. The car was right with them when they left their house. They planned to drive west on Interstate 40 until they got to the turn off for Las Vegas. They would then take Highway 93 north to Vegas and just past Vegas, take the road through Death Valley. Once they got to the major highway that followed the Sierra Nevada Mountains, they would decide whether to take that route or go on across the Mojave Desert and pick up Highway 99 at Bakersfield or Interstate 5. This is the point at which they would think about losing the followers.

It was going to be at least a two day drive, so they decided to stop in Las Vegas for the night. They stopped at the Las Vegas Hilton and hung around the casino for a while after dinner. They decided to just stay on Interstate 15 until they got to Barstow and then head north. It wouldn't make a lot of difference either way.

The tail was still with them, and Eddie thought he had detected a second car switching off with the one who had tailed them from the start of the trip. Both were regular cars, that would ensure that Eddie could lose them when he decided to do so. They did not stop frequently, but every time they did Eddie topped off the fuel tank. As they got closer to Fresno Eddie told the others, "I'm going to get off the main highway on the south side of the city and try to lose them in the desert and agricultural area. We can skirt around the city to either side and pick up the main road again then. If that doesn't work we will still have some options before we get into the mountains proper."

It was getting late in the day and if the plan worked as they hoped, they would arrive at the camp site just about the same time they had on the original trip. With the GPS unit it would not be hard to find the place, even after dark. The girls had included a couple of extra flash lights in the food basket, just in case.

When Eddie left the main highway they noted that both cars had taken the same exit. The idea now was to find someplace that the truck could go that the cars could not follow. They might have to travel a good distance to find what they were looking for. Ideally it would be a place where the other cars would get stuck, such as an irrigated area. Eddie passed up a couple of good possibilities because the area was still a bit too populated for his purposes. He finally took a side road across an irrigation canal and paralleled a citrus grove, which in turn gave way to open range. The terrain was hilly enough that Eddie picked his way around the higher hills.

Both of the following cars had dropped way back and Eddie started looking for some rougher terrain that would make it harder for the cars to navigate. He edged closer to the populated area again, looking for a likely spot to have the cars get stuck trying to follow. He was far enough ahead of them now that he could lose them by simply driving faster over the rough terrain than the cars could manage. As luck would have it, he came to a leaking irrigation line that had softened the earth enough to make it difficult even for the truck to cross. Eddie felt sure that this would slow them enough to make their getaway, even if they found a way around it. He put the truck into four wheel drive and crossed the muddy area. The tire tracks he made filled with water immediately after they passed.

He started looking for a way back to the main road and within fifteen minutes was out of town on the north side, traveling toward Yosemite. Although they kept a sharp lookout for the two cars, they did not detect them, and they looked for other vehicles that might be following as well. Eddie thought they were clear and continued on Highway 41, headed northeast toward Yosemite. When he came to the Highway 49 intersection he took that to the north so that he would be able to retrace their first trip. Although they had the GPS, there were only so many ways you could travel in the national forest. Each of them tried to recollect the route they had taken on that fateful Friday almost a year ago.

It was again dark when they arrived at the location, but Eddie recognized the meadow immediately when he came to it. He drove to the exact same spot they had used on their previous trip. The rocks they had placed for the campfire were still there, blackened by the flames.

After Eddie shut off the engine, they just sat in the truck for a few minutes, each with their own thoughts about the wisdom, or foolishness of their current actions. Finally Bill said, "Well, we're not accomplishing anything by sitting here. Let's get out and see if Joe makes contact, or we have to spend the night."

The four got out of the truck and stood around the old campfire in the light given off by one of the flashlights Beth had taken out of the basket. The silence was ominous. It seemed that no one wanted to break it. Even the nocturnal animals were quiet.

Suddenly, they seemed to drift away, as they had done the first time. Joe's voice said, "Welcome back. And to what do I owe the pleasure of your company so soon?"

"We are a bit confused," Beth said.

"Make that a lot confused," Cindy said.

"And what is the source of your confusion?" Joe asked.

"Too many things don't make sense about what you told us in our earlier encounter," Eddie said. "We would like you to explain a few things to us."

"Okay, what bothers you most?"

Bill took that one. "You said we were to save the world from destruction. We have looked at every way we could conceive that would lead to destruction of the earth and nothing adds up. Could you explain that a little better?"

"No. I was truthful with you about your purpose being to save the earth from destruction. I didn't say how it would be destroyed, and cannot do so. You can rest assured though, that your mission has not changed. You are to save the world from destruction. When the time is right, you will see the truth of what I am telling you."

Eddie added, "We found the proof of the evil part of this equation. Someone has had us followed for a long time, probably since just after our last meeting with you. We have about decided the man behind it is one of the richest men in the world and he has been on the scene for over forty years. The best we can figure is that your counterpart gave him clues as to what to look for to identify anyone you enlisted on your side. What we don't know, and cannot figure out, is the significance of this information, and what his intentions are. If he simply wants to get rid of us, then he could have done that already. We have doubts that you were one hundred percent truthful with us during our first meeting."

"I'll admit that I was not very forthcoming, but I did not intentionally mislead you. I omitted some things that would have no doubt made it easier for you to learn some of the things, but you needed the experience of looking out for yourselves."

"Can you give us an example of that?" Beth asked.

"I knew that my counterpart as you refer to him had someone in place to seek you out. You needed to discover this for yourselves, as you will have to devise some way to deal with him on your own. I told you at the outset that I was simply going to give you the tools, the mental tools, to address the problems."

"One thing that has us all hung up," Bill said, "is the inheritance I received from an Aunt I hardly knew at precisely the time that we had our first session with you. The odds are too great for that to be a coincidence and we were wondering if you could shed some light on that incident?"

"I'm afraid I was a bit overconfident, or more accurately, a bit less impressed initially with the cognitive ability of you four. I admit I made arrangements for the money, but I did not harm your Aunt. As a matter of fact she is living out her final years in a rest home with the best care money can buy, as you folks say. She is suffering from advanced Alzheimer's disease. There is another like you four, who does little chores like that for me from time to time. I had her draw up the will and date it, then put the money into an account in your Aunt's name. Knowing that you would call to check, she waited at the office of the fictitious lawyer until you called and verified the information you wanted to hear. It was a harmless way to get you four started."

Eddie said, "That helps to clear up a few things, but I still have a problem with the other alien. He has had a man in position for over forty years, getting richer every day, but I can't see how this has anything to do with destroying the earth. I don't understand the relationship between you and the other alien. If his purpose is simply to keep us from saving the planet, then all he has to do is have us killed. So far this has not seemed to be his intent. Can you tell us a bit more about the relationship of the two of you?"

"For thousands of years we have had an adversarial relationship. He seeks to restrain the advancement of your civilization, while my task is to nurture the advancement. There have been times in the past when he has had a certain degree of success. I had the Egyptians on the verge of becoming the dominant civilization in the world, but through a single individual he managed to tear down much of what I had built. Then again, I dealt him a devastating blow when I engineered the defeat of Attila the Hun. The same is true of Hitler in times more familiar to you. The ebb and flow of history can tell you whether he or I was victorious. He is in a purely defensive mode at the present time because he knows he will not have to do anything specific to ensure the destruction of the earth except to derail the efforts of you four to prevent a cataclysmic event."

"Could you expand on that a little more?" asked Cindy.

"I cannot say more. You must do this on your own."

"Is this just some sort of game to you two? You have been here ten thousand years playing games with the development of the human race for your own entertainment?" Eddie said.

"We don't look on it as a game, but as our destiny. My destiny is to nurture the development of your kind, while his destiny is to destroy what you manage to build up. We have very strict rules about interference with the natural order of your progression that we both must follow. It is just who and what we are."

"I still see that as a game, whether you call it destiny or superior intelligence, or something else," Beth said.

"If we allowed you access to things that your kind are not prepared for, it could cause problems that you could not deal with. Technological advancement takes place in stages. One thing is invented, which impacts something else, which spurs another invention, sort of like building blocks. You four should be able to see that a world populated with people who would be your peers could have devastating effects on your world. Think of a million new inventions per week to deal with, all well thought out and practical. Your world would be in chaos within a year."

"I'm sorry but I still don't understand this. You say the bad guy won't have to do anything to ensure the earth's destruction but stop us from doing something. If he has his man just sit back and watch us what will that accomplish? We have all looked for some indicator that would even remotely lead to the earth's destruction and we haven't come up with anything. No matter what technology now in existence would be used, nothing would do the job completely, and we deduced that you were talking about total obliteration. Is that the case?"

"Yes, total and complete annihilation."

"And you can't even give us a hint?"

"I have told you as much as I am allowed. You have the ability to do the job; you just have to figure out what the job is."

"And do you have any suggestions about how to deal with Clarence Woodman, who has been having us followed ever step we take? If we take care of the ones he has on the job now, he will just hire others. If we go directly to him with the accusations he will deny them, and legally we don't have a leg to stand on. That will only confirm for him that we are your pawns."

"You will figure something out. I cannot tell you anything other than to fight fire with fire."

Within a matter of seconds they found themselves back at the camp site. Bill said, "Well I am sure glad we got all that settled. I now feel so much better about the situation."

"I didn't expect to get satisfactory answers to some of the questions, but I expected he would be a bit more forthcoming about the destruction of the world thing. What I got out of that conversation was that whatever is going to happen will not be caused by the evil twin, but that he will be trying to thwart us from doing some specific thing that will prevent the cataclysmic event. Is that the way you guys read that?" Beth asked.

"I agree with your interpretation, but I still don't see anything that fits the scenario. Whatever is going to happen seems to be outside the known parameters of anything in existence now," Cindy said.

"At least we got the question of Bill's inheritance cleared up. That was a big stumbling block for me. It also tells us that he knew about Clarence Woodman before he chose us for this experiment, or mission, or whatever we want to call it. I would assume he has had someone watching Woodman just like Woodman has someone watching us. The only difference is that it takes a lot more manpower to keep track of four than one," Eddie said.

"As far as I can see we are back to square one on some issues, while we understand the situation a little better. I am not sure how that helps us, but that's my take," Beth said.

Bill seemed to be deep in thought. Finally he said, "What if we are going about this all wrong? We have been going under the assumption that the world was going to be destroyed by some evil force using weapons of mass destruction now in existence. Joe pretty much implied that will not be the case when he said that all the bad guys have to do is watch and wait to prevent us from doing something that will save the world."

"If that's the case then the people working against us will be as vulnerable to annihilation as we will, and it doesn't seem logical that people would work toward their own destruction willingly. They would have to be ignorant of the outcome, or have been promised immortality or some version of life after the big event," Cindy said.

Eddie replied, "Some people would be dumb enough to do it, but not the numbers that would be required to sidetrack us once we accumulate a great deal of wealth. I believe this session tells us that we have to be on the lookout for things out of the ordinary that might be the cause of such destruction as Joe talked about. I got the impression that he was talking about total annihilation, like with nothing left but the dust."

"It would take something really outside our realm of understanding to cause such an event. Eddie, the other day you made a comment about a giant meteor knocking the earth out of orbit, but that might be the kind of thing we are supposed to be looking for. The impact alone from something like that would probably wipe out half the earth's population in one fell swoop. The rest would have the elements to cope with in a very short period of time," Beth said.

"Suppose it is something like that?" Cindy said. "We would have a limited time to come up with something to counter such an event, and even our superior intelligence would only go so far."

"I think we have to look for something outside our experience to be the catalyst for whatever is going to happen. Don't discount anything that even remotely has possibilities," Bill said.

"Well, we are not going to solve the problem tonight. Let's sleep on it and brainstorm again in the morning," Cindy said.

They erected the tents and bedded down for the night.

Chapter 7

When they awoke the following morning they sat around the campfire and hashed over the happenings of the previous night. They all seemed to understand a little better about the overall situation, but were no closer to solving the major problem than they were before they came here.

"Let me ask you guys a question," Eddie said. "Do you have a different opinion about Joe now? Did what he told us make you look at him in a more favorable light, or was he just blowing smoke to salve our doubts?"

No one answered right away. Each was processing their own thoughts and analyzing their feelings toward the alien entity. He had to know that this would happen, and the probable outcome.

Bill finally replied. "The two things that I was most hung up on had to do with the money thing from my Aunt, and the fact that there was nothing out there that could even remotely foretell the coming destruction of the earth. He dealt with both of those issues to my satisfaction, not that I understand what they are supposed to mean. The hang up I had about the other alien being set on destroying the world still weighs on me somewhat, but his explanation helped some. The alien couldn't care less because he will not be directly affected. He could be using Woodman under false pretenses, which would make more sense, since no one who is not suicidal helps plot their own destruction."

"I sort of feel the same. I have lost the total distrust I had been building up to. What Joe said about world chaos, if everyone was as smart as us, kind of makes sense too. I mean, who would have time to look at what other people are doing if they are busy doing their own thing?" Beth said.

"So where does that leave us?" Cindy asked. "Do we just continue as we are going, trying to build up wealth to act when the time comes? And how are we going to recognize when the right time comes? There are still a lot of unknowns and they are going to have to be dealt with as we discover them."

"What kind of things could cause total destruction of the planet?" Eddie asked. "No matter how off the wall, we have to look at anything that could possibly have that effect. I mentioned the thing about meteors because I saw it in a movie once where a huge asteroid was bearing down on the earth and they had to find a way to obliterate it or change its course. That is one scenario that I think could easily become a reality. We will have to do the math and physics, but it might be a start to determine how big a meteor would have to be to have a cataclysmic effect on the earth. We have astronomers studying those things all the time, and if something is even a possibility we will hear about it."

He continued, "Another possibility that comes to mind is the earth simply exploding from within. We know the gas and elements beneath the earth could produce enough energy to do the trick, but not what could cause such an event."

Bill said, "The earth could somehow change its orbit, but unless it were dramatic, there's no way an event such as Joe described could happen. Diseases are out too, since he said total destruction. Other than the development of weapons unknown to us now, there are not many other possibilities."

Beth continued the thought. "I'm sure glad we came up here. All our problems are solved now; we just have to discover how to deal with the solutions."

"I do feel better about it," Cindy said. "Not the nice warm feeling, but the kind that a better understanding of the situation brings. I am not as concerned about the people watching us now, although I think we should do something about that. The dilemma about being torn between working on money projects or the world destruction thing is a load off the mind. Knowing for sure about the money Bill got helps too."

"I can't see us doing much different than we have been. I now have a feeling that the thought computer idea was implanted in my mind to move me in that direction. I don't have any idea about the implication of that, but it seems more important than some of the other stuff we are working on. Maybe it is supposed to come into play in some way with the main project," Eddie said.

"That is not out of the realm of possibility. I guess if the alien can do what he did to all of us, then it would be a simple matter to direct our efforts in any manner they chose," Bill said.

"Are there any other possibilities any of you can think of?"Eddie asked.

"What about the sun burning a hole in the ozone layer?" Cindy asked, "Could that be catastrophic enough to cause total destruction?"

"I wouldn't discount it totally, but I don't believe it would do the kind of destruction Joe seemed to be hinting at immediately," Bill said.

"I still think we are back to watching and waiting for some indicator. Have you guys given any thought to the premise that maybe we are not meant to participate in the culmination of the act, but something we invent will be what is used to prevent the calamity?" Eddie asked.

Again Bill answered first. "That could be, but it doesn't feel right to me. Somehow it seems more urgent than that. I have no basis for the feeling, just that it is something we are supposed to do ourselves."

"I agree with Bill," Beth said. "How I long for the days when picking out the clothes I was going to wear that day was the biggest decision I had to make!"

"I guess it wasn't a totally wasted trip. We all seem to have a better feeling about our situation now," Eddie said.

"I don't know if it is better or just different, but I do believe we will be better able to deal with things after this session," Cindy said.

"So what do we want to do about the people who have been watching us? Do we let them continue, or confront them and tell them to go back where they came from?" Eddie asked.

"I am thinking that maybe a call to Woodman might be in order," Beth said. "If we just tell him flat out that we are the people he has been looking for and dare him to do anything about it, then he would no longer have a reason to have us followed. We might suggest that it could be beneficial to sit down and have a little talk."

"That idea has merit, but I don't think it is the right time to do it now. Maybe somewhere down the road it could become necessary," Bill said.

"I say we continue with the stuff we are working on now and keep a wary eye on the people watching us, until something happens to make us want to change the situation," Eddie said. "They must know that we can take care of them any time we want, or maybe they don't and ignorance is bliss."

"It isn't going to do us any good to worry about it. We will just have to take it one day at a time and be on the lookout for anything that has possibilities for calamity," Beth said.

The four spent the remainder of the day relaxing and talking more about their situation. They exhausted all the thoughts they could summon that even remotely related to the problem. Their total worth with the payment from the software company Eddie had helped with the processor was in excess of two million dollars, though a lot was invested. The stock had already started to climb and that investment alone would be worth another couple of million within six months. Financially they were secure, but man was not meant to sit around doing nothing, and their energy would result in the accumulation of vast amounts of money.

They spent a lot of time sitting around the campsite quietly, each contemplating their situation and how to deal with the different aspects. The people watching them were a concern, but not a great one. They knew where they hung out, and how many there were. They would catalog what they looked like and possibly even try to get a surveillance system inside their warehouse. It couldn't hurt to keep a closer eye on them.

Woodman was the enigma in the entire situation. He couldn't know about the coming destruction. It was beyond any of them to imagine that he could participate in such a scenario, knowing that he would perish with everyone else. True, the other alien could have convinced him that he could protect him and whisk him away from the calamity, but Woodman had certainly been given the same intelligence capacity that they had, and his reason would not allow him to accept that as truth. He had lived a life of luxury for over forty years, and possibly didn't even know the reason for it.

It was not beyond reason that he had simply been told to look for someone like himself to appear as the sole reason for his encounter. If that was the case, then he would have to make contact with his alien to get instructions. That made more sense than any other proposition the alien could have presented to him.

Though they did not verbalize a lot that Saturday afternoon, each knew that the others were trying to better understand all the nuances of their situation. None experienced any degree of satisfaction from the ruminations.

On Sunday morning they got up and had breakfast, broke camp and headed back down from the mountains. The two day drive back to Albuquerque was uneventful and they did not see anything of the people who had been following them.

When they entered the offices on Tuesday morning it was obvious to all that someone had been in the offices during their absence. "Did anyone have anything accessible that they might have gotten," Eddie asked.

From the start they had encrypted all the data in their computers and they assumed that no one had accessed that. The financial documents were not encrypted but were password protected. That would not be difficult for a really savvy computer expert to crack, but it would not be that easy either. They didn't touch anything until all of them had looked around thoroughly to get a feel for the focus of the break-in. None could see that anything was missing, though they had not gone through the file cabinets carefully. The opinion was unanimous that the focus had been on the computers.

The intruders had been professional because they left no evidence of their entry, and had not the occupants been extra sensitive they might not have noticed anything amiss.

"Do we want to call the police, or look into this on our own?" Eddie asked.

"I see no advantage to calling the police. We might dust for fingerprints and lift those. We can hack into the government fingerprint files and see if we can match anything up. Even if we do it will not provide a great advantage. We know who the guilty parties are, and can confront them at any time. It would be nice to know what they were searching for, or if it was just a fishing expedition," Cindy said.

"I believe it is time for us to return the favor to them. Give me a couple of days to design the equipment and we will lure them away from their warehouse and set up a surveillance system inside. That way we can keep track of the numbers and if I can get a good sound pick up, we will know what they talk about," Eddie said.

"That's a good idea. I don't think they will suspect us of doing something like that. I get the impression that these guys are petty crooks and not really too sophisticated," Beth said.

"We have to figure out a way to get them all out of the warehouse at the same time. Maybe something like we did before. I have to think that they always leave at least one person on duty there. We will have to devise some ruse to get them all away at once," Eddie said.

"You get the equipment designed and we will find a way to get them all out of the building. In the meantime, let's work as usual. I have everything running smoothly for the FDA tests. Beth and Cindy are going to operate the web site for the chip change. You know we are going to make an awful lot of money off that," Bill said.

"Yes," Eddie said, "everyone with a computer is going to want the upgrade, but I expect your AIDS drug will do even better than that."

"Eddie, are you still hung up on the computer responses to the brain impulses?" Cindy asked.

"Yeah, I am having a hard time figuring out how to get the impulses from the computer back to the brain in a form that the brain can recognize and interpret them instantaneously. The impulses travel so much faster from the brain that it is hard to get the computer to react as quickly. I have a feeling that the return signal is going to have to be sequenced in a manner similar to what the brain produces, except on a different frequency. I need to do an in-depth study of the signals the brain puts out to see exactly how the brain produces the impulses. It might be that the brain uses different notes or harmonics, much like the notes of a song. I don't know how many permutations there would be, but it has to be an astronomical number."

He continued, "When someone speaks to you, your ear hears the sounds and interprets them to the brain. The sounds must be within a certain frequency range for the human ear to hear. The brain communicates with the ear and interprets what those sounds mean. My idea is to make the computer the ear and have the impulses generated by the computer in that frequency range, like the human ear, that the brain can interpret. The principle is the same, only the medium is different. The ear is integrated into the brains function and operates as a seamless organism. I use the electrodes from the computer in much the same way the brain uses the ear. I just have to find a frequency that the brain will recognize like it does with the ear. I have a feeling that it is going to be in the upper frequency spectrum because the higher frequencies can carry more data, and much faster than the lower frequencies."

"I would offer to try to help, but that is way over my head, even with my expanded intelligence," Beth said.

"I don't know why I am so obsessed with the idea, but it has been that way from the first time we met Joe. It wouldn't surprise me if we find this is tied somehow to what we are supposed to be doing in the save the world scenario," Eddie said.

"Is there any way we can help?" Bill asked.

"I don't know how, unless you could come up with something to control the frequencies at which the computer sends the signal back to the electrodes attached to the head of the user."

"Couldn't you use some sort of oscilloscope to measure the frequencies and maybe put in an intermediate booster or reducer to control the frequencies?" Bill asked.

"Possibly. The key is finding the frequencies that the brain process, and if we had something like that I could experiment until I identified the frequency range. I could then modulate the impulses to change the frequency. I hadn't thought of trying that, and maybe you are onto something," Eddie said.

Eddie and Bill went to Eddie's work space. Eddie looked for the materials he would need to implement the suggestion Bill had made. He found the micro thin wire he needed and attached it to each of the leads from the computer to the helmet like device which contained the electrodes that handled the communication link between the computer and the brain and vice versa. He placed the device on his head and powered up the computer. He asked Bill to observe the frequencies of the impulses from both directions.

Bill did so and noted the figures on a piece of graph paper. The signals from the brain were like a burst of static and he tried to get a finer measurement by manipulating the bandwidth of the impulses. He was able to refine the signal somewhat, but there was so much data on each impulse that even expanding the signal to the maximum allowed by the constraints of the oscilloscope the signal was still too cluttered to discern individual characteristics of the data elements.

The computer seemed to be reading the brainwaves because it was responding to the stimuli that Eddie was providing. The return signal however, was not as heavy laden with data points as the incoming signal.

"The stream coming back is not as robust as your brain waves going into the computer. Is there any way to speed up the response?"

"I don't think it makes a difference. The ear processes sound at different data rates and the brain interprets those. When a person speaks slowly the brain processes that, and when the speech is faster, it also processes that. On that basis it would seem that the rapidity of the computer response would not matter. The fact is that my brain cannot process what is coming back from the computer. Be sure you get the frequencies of my brain waves and we will put you under the helmet and see if the person makes any differences," Eddie said.

The two switched places and Bill had the electrodes in contact with his head. When Eddie turned on the computer Bill asked, "What am I supposed to do?"

"Just look at the screen and visualize what you want the computer to show you," Eddie told him.

Bill did as instructed and Eddie checked the oscilloscope reading against what Bill had written down for him. To his surprise, the frequencies were very different. They were in the same spectrum, but different.

"Well, this is something I hadn't counted on. It would appear that our brains operate on different frequencies for the outgoing impulses. The computer is still reading something, because it is trying to communicate back to you. I need to try this with the girls too. It may be that our brains are like our fingerprints, none alike."

"What difference will that make to the return signal from the computer?" Bill asked.

"I don't rightly know. It could mean that the return signal has to be on a discrete frequency that is different for each person, in which case this idea will not work. It would be cost prohibitive to have to build each computer as a customized model that would work with only a single person. That would be great for the upper-crust of society, but the poor people couldn't afford them."

"There has to be a way to have the computer recognize the frequency of the impulses from the person and know what frequencies the response has to be on," Bill said.

"There probably is, but in order to design something to do that, I need to isolate one instance where the impulse from the computer can be read by the brain. I might be able to deduce from that the harmonics and logic of the circuit. I tend to think that the differences will form a pattern I can exploit. If I can measure the difference in the frequencies for one case, then I can apply that data to the next case and go up or down until I hit on the second, then the third, and so on. I am betting that the solution will be like a cryptographic key. Once you find that key, then the situation keeps repeating itself ad-infinitum.

There may be as few as a dozen discrete frequencies, or there may be thousands. The practicality of the end product will depend on that number."

"What you are saying is that if there are a large number of variants, then you will have to build multiple versions and have the prospective customer try out computers until they find one that works with their brain impulses, like trying on a pair of shoes," Bill said.

"That's a very good analogy," Eddie said. "It is possible to build in a processor of some sort that can recognize the impulse frequencies and have the computer respond to whatever the values are for that data set. I believe it is workable, but I need to see how many variants of the brain impulses there are, and find a response for one of them."

"How different are the impulse frequencies from the brain to the ones coming out of the computer right now?" Bill asked.

"There is a good deal of difference, because I did not factor that into the first part of the project. I was concentrating on getting the computer to read the brain impulses. That apparently works, because the computer is trying to talk back."

"Let me set the computer up to step through the harmonics and hook up to an oscilloscope. Maybe I can isolate the frequency I need that way."

Eddie spent a few minutes making the changes and connections he needed. He then said to Bill, "Let's try it again."

Bill donned the head set again and Eddie activated the computer. As he watched the oscilloscope he thought he detected a spike. "Did you feel anything just now?" he asked Bill.

"I don't think so," was the reply.

"I'm going to slow down the stepping rate. It is in a frequency agile mode, so it steps through the frequencies automatically until it finds one your brain can copy. I thought I saw a blip just now, and if so that would indicate that you should be getting two way communications with the computer."

Once the adjustments were made he turned the computer on again. Bill said, "I think I felt something then."

"Okay, I am going to set on that discrete frequency and try it."

He made the adjustments and turned it on again. Bill's face lit up, giving him the answer to his next question. "I assume we hit pay dirt."

"Boy, did you ever. This is amazing. The computer is responding so quickly that I can't even digest the responses."

"Well, that tells me that the response has to be on a discrete frequency, but I still don't know how many there will be. Get Beth and Cindy in and let's see if they are different from you."

Bill got the girls and Eddie had Beth don the headset. "Let me know if you feel anything like a response."

He had set the system for frequency agile operation and the blip showed momentarily just as it had with Bill, but on a different frequency. Beth did not give any indication that she had felt anything.

"I got a hit, so I am going to slow down to half speed. You should be able to feel it this time." He made the adjustments and turned the computer on again.

Beth said, "I think I felt something, but I am not sure."

"I'm going to set it on the frequency that should be able to communicate with you now."

He adjusted to the frequency, which was a harmonic of the one Bill had been compatible with, and turned the computer on again. The result was pretty much the same as with Bill.

"Wow, this thing is so fast I can't keep up with it, even in my thoughts."

"That's because you are using so much of your mental capacity. You have to remember that the population at large will not react as you do. They will need to take the time to dwell on the computer responses and their thoughts will not continue until they have digested what the computer is telling them. At least conceptually that is what happens."

Cindy took her turn and the response had to be on a different frequency for her, but with the oscilloscope and frequency agility, it was a piece of cake. Even if the process had to be carried out for every computer sale, they could still process large numbers, and once Eddie got all the frequencies, he could program the computer to look for the spike and then go to continuous mode of operation on that frequency. He felt he had a handle on it now.

He showed Bill what to do with the oscilloscope and donned the headset himself. Once they got to his frequency and the interface initiated, Eddie was just as enthralled as the others had been. Talk about a revolution; this was so far beyond what was available now that it was like going from row boats to ocean liners, or bi-planes to jet aircraft.

He experimented a bit, trying to slow his thought process enough to digest the data the computer was providing. He was marginally successful, but didn't consider this very important in the overall scheme of things. It was not that important for the four of them to be able to use the machines. They already had the capacity to do pretty much anything they wanted to. The only important thing to them was being able to access the combined pool of their knowledge about the earth's destruction. He would find a way to do that after he ironed out all the bugs.

Bill said, "Now that you know it works, how long before you can go to production?"

"Not for a while yet. I still need to isolate all the harmonics that the brain can use, based on the outputs. I am going to need some outsiders to complete that phase of it. The participants will not even know what they are doing. I will keep the screen turned away from them and isolate the frequencies from the oscilloscope."

"How many test subjects will you need?" Cindy asked.

"That's an unknown at the present time. I will have to get a large enough sample to be sure I cover the entire spectrum the brain is capable of interfacing with. I really don't think it will be over 20 to 50. The frequencies for the four of us were all harmonics of the frequency our brains transmit on. The same should hold true for everyone, with the possible exception of people who are autistic, or have some other malady that affects the brain."

"Are you close enough that we should start thinking about manufacturing and such?" Beth asked.

"I believe we can farm the production out to other companies, for both components. There is nothing that can be learned from the wiring in the computer itself. All they will be producing is an enclosure with wiring of a specified size going in one side and out the other. The visual monitor will operate on a free electron flow from the wiring to the display. The interface unit we will produce in house. It is not hard, and once I show you guys how, we can probably build two or three hundred a day with just the four of us. We can also hire people to put them together in the final step of the process. I still need to work out all the details, but I know I will farm out the production of all the elements except the actual coding of the headset and the computer/monitor interface."

"This is probably a dumb question, but can someone possibly reverse engineer the system?" Cindy asked.

"It isn't a dumb question at all. That is one of the first things I thought about when I started with this project. I have traps and pitfalls in every step of the process that will counter any hacking technique I know about, and probably some that I made up as I went along. One false step destroys the entire package, and each user will be warned of the consequences of tampering with anything about the system. That will be hard for a lot of people to swallow, especially those who make their living in the computer world."

Beth said, "Not to change the subject, but I had a thought about how we could get the bugging camera installed to watch our watchers. I don't know how practical it is, but what if we lure all of them away but one, and tranquilize him long enough to install the camera?"

"How do you propose we do that?" Eddie asked.

"With a dart gun like they use to tranquilize wild animals. We will have to probably concoct our own tranquilizer, but I think it might be workable."

Bill said, "I think she may be onto something. I can search for the right combination of drugs to get something fast acting, and with an amnesiac effect. If we can put him to sleep without being seen, he probably will not remember how he got to wherever he is when he wakes up."

"How are we going to lure him away from the monitor long enough to install a camera? We will need between half an hour and an hour, and I don't see any way to zap him without being seen," Eddie said.

"Simple," said Beth. "We rent a scissor lift, and after we have lured everyone else away from the building, you just drive up to the side of the building and raise the lift high enough to step onto the roof. You approach their camera from a blind side and put a bag over it. He will have to come up to check it out, and when he does you zap him. You can then do your thing, remove the cover from his camera, return to the lift and leave the same way you came."

"It sounds too simple to work, but I think you have hit upon a solution," Eddie said. "Bill, can you come up with whatever we need to put the guy out?"

"I don't think it will be a problem. If some of the drugs I need are controlled, I will get one of the doctors on my project to write a prescription. I think I can probably have the concoction ready in twenty four hours."

Cindy said, "I will look for the tranquilizer gun and dart syringes. They shouldn't be too hard to come by, but I might have to order them, so factor some time for that into the plan."

Everything they needed was on hand by the following evening. They planned to try to execute the plan the next day. Eddie called the scissor lift company and told them he wanted to rent a lift, trailer, and truck, all put together and ready to go by noon on the following day. He was assured that everything would be ready.

Bill told Eddie, "I mixed this for about an hour, but the time the guy is out will depend on his body weight. I figured for 180 pounds, so if he is smaller, he will sleep longer, but if he is larger, then you will have less time, so try to work fast."

Beth drove her car around to the side of the building away from the camera and Eddie went out the back door and got into the back seat of her car and covered himself with a blanket. She then drove to within a block of the rental company and turned a corner to shield Eddie's exit from the car, and he quickly jumped out and secreted himself in a doorway. The tailing car did not notice and continued tailing Beth.

Eddie called Bill and told him that he was safely out of the car and was on his way to the rental yard. Cindy would leave now and when Eddie got in place with the truck and trailer, Bill would exit.

The entire ploy depended on their only being four people on the surveillance team.

When Eddie got within a block and a half of the road from which the watcher would exit, he called Bill and told him where he was.

Bill soon drove away from the building and Eddie watched as another car set out in pursuit. He gave it a bit more time and drove the rented truck alongside the building where the cars had exited the garage. He then raised the scissor lift and stepped gently onto the roof. He then very quietly made his way to the structure on which the camera was mounted and approaching from the rear, slid the bag over the camera. He then took up a position opposite from the way the man would have to turn to access the camera.

Eddie could hear the steps creaking with the weight of the man ascending the stairs. The door to the roof opened, and as the man stepped out, Eddie shot him in the right buttock with the dart syringe. He then quickly ducked back behind the wall out of sight.

Bill had told him that it would take the drugs from ten to thirty seconds to take effect, and Eddie slid on around behind the shed in case the man came his way. Pretty soon he heard a thump as the man hit the roof. He still waited a full minute before going to check on the man.

He had managed to extract the syringe from his buttock, but was still within a step of where Eddie had shot him. Eddie retrieved the syringe from the man's hand and then massaged the spot where he had shot the man. He didn't know if the massage would help, but it couldn't hurt. He checked the man's eyes for any reaction and quickly went down the stairs.

He had the camera in a Musset bag over his shoulder and looked around for the most likely position to install it. The camera itself was about the size of a small button and was an off white in color, hopefully to match the color of the walls. The sending unit with the battery pack would be mounted out of sight above the plaster, and the only thing showing would be the small button like camera.

They would have to be looking for it to see it. Eddie managed to find a place to get the battery pack and transmitter out of sight near enough to the camera location to be workable.

By the time the installation was done, he had been in the building for just over thirty minutes. He double checked to make sure everything was operational, then removed all evidence of his presence. He returned to the roof and checked the man again. He was still out, so he removed the bag he had placed over the camera, left the man where he lay and retraced his steps to the scissor lift. He lowered it back to the travel position, got in the truck and headed back to the rental yard.

The man was going to wonder at the short length of time he had kept a piece of equipment that he had paid for the entire day, but that couldn't be helped.

After he had returned the equipment, he called Beth and told her that he was ready to be picked up at the same place. He also called Cindy and asked her to call Bill and tell him that the plan had been a success so far. It would depend on what the man on the roof remembered as to whether the plan was entirely successful.

Beth picked him up and he returned to their warehouse the same way he had left. Soon they were all back in the building and Eddie said, "Well, let's see how successful we were." He turned the monitor on that had been hooked to the camera system and was rewarded with a picture of the man he had tranquilized sitting at a desk in front of the monitor showing the door to their building, with a blank look on his face. The other three men sat on couches and appeared to be conversing with the man behind the desk.

Everything seemed to be in order, so they assumed that the man shot with the dart did not remember what had happened.

"Now we will know what they are doing, but that's not going to help us much in deciding how to deal with them," Cindy said.

"We won't have to worry about it for a good while. Let's just keep an eye on them while we are taking care of business. I don't want to do anything until we see how they react to the introduction of the new computer, and that's not going to be for several months yet," Eddie said.

"What we really need to do is figure out some way to approach their boss and confront him about their activities," Beth said. "I am still a bit miffed about the way they abducted me."

"We can work on a plan for that, but I really believe the most important thing for us to do now is work on the end of the world scenario. If the computer Eddie has built doesn't give us some clue, then I am at a total loss," Bill said.

"What's the time line on the drug, Bill?" asked Eddie.

"The initial trials will be completed in a couple of weeks. We then have to review all the test data with the FDA, and if they approve everything, then it can hit the market within a couple of months after that. I have already been getting feelers from the big drug companies. Word has spread about the test program, and I don't doubt that some of the personnel involved have been keeping them up to date on the progress and prognosis for success."

"Are you still thinking about selling the formulation outright, or do you want to hold onto the patent and just contract for the production process?" Eddie asked.

"In the beginning I thought that we wanted to stay under the radar about the huge amounts of money we will be amassing, but I see no point to that any longer, so I think we should patent the drug and contract the production. The production process is going to be fairly reasonable, and if we don't look to make huge amounts of money, we can sell the product at a cost that insures the people who really need it most, namely the poor, can afford it."

"I tend to agree. We know who our opponent is, and it will be a matter of outsmarting him. He already suspects that we are who he has been looking for, and as we gain greater recognition for the things we are doing, it will make it harder for him to act overtly against us," Eddie said.

"I say we just continue the way we are going and worry about him later," Cindy said.

"That's about all we can do until we are ready to confront him, which we are going to have to do sooner or later," Beth added.

"I am about ready to call it a day," Eddie said.

Everyone agreed and they returned to the house they were sharing.

Chapter 8

As the days passed, each of the four worked long hours. Beth and Cindy were kept busy with the computer upgrades, which had taken the market by storm. It was fortunate that they had automated the procedure; otherwise they could not have kept pace with the orders.

Their wealth was growing by leaps and bounds, and they did not even have the time to look for other ways to invest the money.

Bill's AIDS drug had completed the testing and was now being reviewed by the FDA, in conjunction with Bill and some of his doctors. He had applied for the patent and selected a drug company to produce the drug. He had also contacted several humanitarian organizations looking for ways to get the drug to the impoverished countries where it was needed most but was unlikely to be available without some sort of subsidy. The drug was getting a lot of headlines as one of the most remarkable advances in medicine for the past hundred years.

The drug would be made available to the organizations Bill chose to distribute it for the cost of production. This assured that the FDA would be under pressure to do the review forthwith. Bill had been offered 200 million dollars for the patent on the drug, but shunned the offer.

In the meantime, Eddie had gone to the streets for subjects to test for his computer project. He paid participants a hundred dollars and simply went to a different section of the city each time and got five people to test.

He verified his presumption that the brain waves were all within a given frequency range, and that the responses coming from the computer were all harmonics of the brain frequencies. The total list of possibilities was 25 squared. He programmed the 25 response frequencies into a frequency agile mode and no matter what frequency the brain transmitted on, one of the harmonics in the program worked for the return signal.

Eddie drew up the specifications for the hardware and found a company to start producing the computer towers. Since they would have to retool the production line, the initial cost was pretty high, so to ensure the company that it was not a onetime order, he set the initial number of units at half a million.

The cost per unit was less than twenty dollars, and it looked like he would be better off setting up a factory to build the headset interface. He had started looking for a location where the labor market was not exorbitant and building costs were reasonable. He finally chose a location near Atlanta, Georgia, and negotiated some lucrative tax incentives before he bought the property.

Once that was done, he hired an architect to do the design in accordance with the drawings he provided. It would take an estimated six months to complete the structure from the time they broke ground. He hired an engineering firm to do the advanced work and they were busy while the property acquisition and permits process were taking place.

Because he wanted to use the first computer for the group's own purposes, Eddie built a second prototype to do all the testing and design work.

Within six months they were ready to start the production process for the hardware. Since they needed someplace to store the product, they decided to do the final phase of the process right in Albuquerque. Eddie figured they were about a year away from the final programming phase and asked Beth and Cindy to find a good location in the area and start looking for a builder to design and build what they wanted. This would be a state of the art facility, both in design and security features. When it was completed, they would move the entire operation for all their activities to it.

Cindy did the office/laboratory portion on a Computer Aided Design (CAD) program, knowing what they had and extrapolating to their projected needs. Beth meanwhile, worked with Eddie to determine how the production line would be set up, and what they needed in the way of robotics to make the job easier.

The facility was planned right down to the number of parking spaces needed, and loading docks and storage facilities for components and finished products. The cost was going to be staggering, but the way the money was rolling in would not cause a strain on their finances.

The computer upgrades, which would be obsolete the minute Eddie's new design hit the market, were making them over twenty million dollars a week, and the other endeavors of Cindy and Beth were steadily increasing in popularity and sales.

Bill had approval for the drug, and the first batch should hit the market within the next two weeks.

All in all, the group was healthy and wealthy. They had no need for financing outside their own sources, and were not constrained by stockholders, so when they made a decision, it was implemented immediately.

In every case where building something was required, they structured the contracts with provisions for penalties and rewards. If the project came in early, then the builder got a bonus. If it was late there was a penalty. Shoddy workmanship was not tolerated, and in some cases builders had to tear out and rebuild.

Eddie and Bill designed the security system for the new headquarters building and processing facility in Albuquerque, and personally fabricated the components for the contractor. There was only one set of the plans, other than the ones Eddie and Bill kept. The supervisor was cautioned that he should not let the papers out of his possession on threat of termination of his contract.

One of the group was at the site all during the construction process, looking for flaws in the design or layout. Several changes were made when they noted a better way of doing something, even eating the cost of the changes.

Their watchers had increased to six, and they probably had someone watching the various projects they had going on in other areas too.

Eddie experimented with the machine they were going to use for their own purposes, and finally managed to slow the process down enough to allow them to dwell on a subject longer. Even with the ability to do that, they could not find any clue to the riddle.

They continued to puzzle over the problem for the better part of a year. Finally Eddie said, "I don't think there is a solution to the problem at this time. We have studied every theory any of us have run across, no matter how absurd it sounded, and we have exhausted all our own thoughts, all without any positive results. Whatever is to happen is not in our knowledge banks yet is the only conclusion I can draw."

Bill said, "I tend to agree. I believe we have exhausted every avenue open to us, so the logical conclusion is that whatever is to happen is beyond our grasp at the present time."

"The only solution I see," Cindy said, "Is to keep on our toes looking for something in the future."

"Yeah," Beth said, "And it could slip by us because we don't know what we are looking for."

"Well, your new computer is going to make a lot of people smarter by a large measure, and maybe one of them will come up with something that will be capable of annihilating the earth," Bill said.

"It could be," Eddie replied. "There's no more we can do about it now, so we might as well put it to rest and get on with making money."

The first of the hardware units was scheduled to be delivered in less than six weeks, and the new building completion was three months away. They decided to use the warehouse they were currently in to store them until the new facility was completed.

The chip sales and programming had brought in close to a hundred million dollars during the previous year. A goodly portion of it had been invested in the facility in Georgia and the one in Albuquerque. With the income from the AIDS drug now being added to the total daily, their worth never got below a hundred million, even with the expenses of accelerated time tables on the projects.

Eddie was chagrined with himself every time he looked at the picture of the people in the warehouse across the way, for not putting an audio pickup into the system. He didn't think the watchers would give up any earth shattering information, but they would probably verify the identity of their employer via phone conversations.

Eddie had made several trips to Atlanta and to the construction project site, to make sure it was on schedule and that there were no changes necessary from the original design. He also set up an employment agency to start lining up employees to operate the plant when it was completed. He also needed to design a training program for prospective employees.

He decided that Beth would be the best to handle that aspect. She seemed to have a better grasp of the little things associated with a task than the others. He would have to give her all the particulars for the process and she could then set up something to train the supervisors, which would have to be identified pretty soon. There was a lot of work that had to be done, but they all seemed to gravitate to different tasks without missing a beat.

Bill had pretty much assumed the task of overseeing the construction in Albuquerque and Eddie the Georgia plant. When Eddie had been negotiating with the Software Company about the chip for the enhanced computer, he had taken a liking to one of the engineers. He contacted him and offered him the job of running the plant in Georgia. He offered a salary that was twice what he was currently making and the two met at the plant to talk it over.

Keith was the man's name and he was in his late twenties. He didn't have a lot of management experience, but was a good people person. Eddie told him that if he accepted the job he would be in charge of selecting the crews and setting up work schedules. Eddie planned to hire a good accounting specialist to handle the financial end of things, and let him staff his own office.

Eddie went over the equipment specifications with Keith and explained that the component he would be building had to interface with what was being produced elsewhere, and that the final mating of the units would be done in Albuquerque.

Eddie told Keith, "The design is something you are not familiar with, and I will need to bring you up to speed on the specifications. The new system will be based on a free electron flow directly to the user's brain. What you are going to be producing is the interface between the computer and the brain. Each of the electrodes has to be precise in order for the system to function. The computer will operate at the speed of thought. The instant a person has a thought; the computer flashes the response on the screen. There will be no need for an internet connection unless the subject is so esoteric that it could not be included in the data package. We have not told anyone about this product because it is so revolutionary that all other computers will become immediately obsolete. Inside of three years, it will be rare to find an ISP still in business."

Keith looked at Eddie with amazement. If he had not been aware of Eddie's involvement with the chip project his company had come out with, he would not have believed him. He still had to stretch his imagination to take him seriously. "That is unbelievable. I have to see this to believe it."

"I will take you to Albuquerque and demonstrate the prototype for you. You have to understand where you fit into the scheme of things in order to do your job, but the beauty of the system is that none of the components will mean anything to anyone until they are mated and activated. Only myself and three others will be capable of doing that. Once the units are ready for use, any tampering with the operating system will destroy the entire system. If any of the components produced here go bad, or are damaged, they will have to be returned to you for repair."

"That really gives you a corner on the market, doesn't it?"

"Yes it does, but that is not the primary reason it is set up that way. The fact is that only four people on the face of the earth can program the units, but the reliability will be such that the failure rate will be only thousandth's of a percent."

"This sounds too good to be true, even knowing your capability. I have a hard time believing something like this exists."

"If you are interested in my proposal, I will take you back to Albuquerque and demonstrate it for you."

"I am definitely interested in the job. I just hope I can live up to your expectations."

"I won't just throw you into the water without a life vest. I plan to work very closely with you until you become confident that you can handle it."

"Then I guess I accept," Keith said.

The two flew back to Albuquerque and Eddie took Keith to their warehouse office. After introducing him to the others, Eddie showed Keith the computer set up. He handed the headpiece to Keith and told him to look at the computer screen and think of what he wanted to see.

"It will take the system a moment to isolate your brainwaves so that it knows what frequency to respond on so your brain can read the return signal." Eddie then turned the computer on.

Keith was totally blown away. "Man, I can't believe this. Even seeing it and experiencing it, I still can't believe it. I can see why you think this will revolutionize the computer industry. The stuff we are using now is junk compared to this, and the speed is just amazing. The second I think of something the answer appears. How does it operate?"

"The entire system is based on free flowing electrons. The brain emits electrical impulses that the computer reads, and the computer answers with impulses that the brain can read. It's kind of like a conversation taking place in overdrive. The system is limited only by the speed at which the human mind can produce thoughts."

"What's to keep people from reverse engineering one of these?" Keith asked.

"I have implanted so many roadblocks that it would take something like one in fifty billion attempts before anyone could get around all of them. And, if they hit even one of the trip wires, the entire system destructs. Having the components will not do anyone any good without the programming capability, and there are no manuals for that."

"You do realize that you are about to become the richest man on earth, and probably one of the most powerful, do you not?"

"I already have more money that I could ever need. We made over a hundred million on the computer upgrade I helped your old company with. And Bill developed a cure for AIDS, which he is practically giving away. This whole project isn't about money, although that is certainly an incentive. It's just pushing the technological envelope as far as we can and watching the results."

"I am definitely in if you still want me. I would like to be able to tell my grandkids that I had a hand in something like this," Keith said.

"Keep this under your hat until we get ready to go public. That probably will not be for quite some time. Your plant needs to be up and running before we do that."

"It's going to be hard to know that something like this exists and not be able to tell anyone about it."

"Do you remember the meeting I had with your old company before they hired me?"

"Yes."

"The president asked me why I was willing to work for so little when I could patent the chip on my own and make much more money. I told him that I was working on something that would make the chip obsolete in three years. This was in the conceptual stage at that time."

"You were that sure it was going to work?"

"I was confident that the concept was sound. I just had to spend a lot of time putting all the pieces together."

"Well, you certainly knew what you were talking about. How did you come up with the idea in the first place?"

"You wouldn't believe me if I told you," Eddie said.

"When do you want me to start to work?"

"Just as soon as possible. Do you want a written contract?"

"I can't imagine why you would want to stiff me with what I have just seen. I will take your word as a contract," Keith said.

"There will probably be a lot more money down the road. We aren't sure exactly what else we will get into, but you will have other opportunities if you are flexible."

"You have already given me more responsibility than I ever thought I would have. It's going to be a real pleasure working for you."

"I will inform the employment agency in Atlanta that you are the company point of contact. Take whatever actions you feel are necessary to get things up and running. You have the authority to commit funds for anything to do with the plant. If you see some flaw during the construction process, make the necessary changes, even if it means additional costs. If something looks like it is going to be cumbersome to operate, try to fix it before it is installed. We tried to use off the shelf materials, but if something looks like it will not interface well, let me know and we will design or buy something that gets the job done."

"Just so I understand, the interface for the headpiece is what we will be making at the plant?"

"Yes, and it is not as complicated as it looks. I will draw up the specifications for each part, and then the plan for mating the components. I will spell out the tolerances for each part, and the quality control has to be ironclad for that part of the process, so don't skimp on the salary for the QC inspectors."

"I can see that you and I are going to have a lot of conversations," Keith said.

"Funny you should mention that. I already have you a cell phone programmed with all the numbers, and a printed directory for who they belong to."

"You must have been sure I would take your offer?"

"Anyone who could see what you just saw and refuse to become involved wouldn't be playing with a full deck. So yes, I was pretty confident that you would accept."

"I am not working on anything that can't be passed on, so it is just a matter of tendering my resignation and cleaning out my desk. I can be back in Atlanta by day after tomorrow."

"I am going to give you a company credit card. It is also a debit card and you can draw or use it for anything connected with company expenses. That includes a car. Look around and lease or buy whatever you will need. I would suggest a four wheel drive, but do whatever you are comfortable with. If you run into any problem with the card, just call the number in ink on the back and the problem will be cleared up instantly."

"Also, before you head back, I want to show you where our new offices will be, also the factory where the final assembly will take place. It's here in Albuquerque."

"Can we do that after I go clean my desk out? I want some time to think about what I have seen, and how I am going to approach the job you have given me. I probably will have a ton of questions that you can answer better face to face than over the phone."

"Whatever works best for you? I will run you to the airport. We, my partners and I, have talked about either leasing or purchasing a small business jet since we are obviously going to be doing a lot of traveling in the near term. I can pick up some brochures while I'm out there."

Chapter 9

The next few months were very busy for all four of them. Keith turned out to be very good in the decision making area, and didn't need a lot of day to day supervision. He knew what was needed and did his best to provide it. He personally screened each prospective employee the agency recommended and had background checks run on every one he anticipated hiring. He did this through a local agency on a set fee per person. The checks were a lot more thorough than the usual police checks, which the agency run, but they also went to the neighborhood and talked to other people who knew the prospective employees.

The salary the company offered for the line work was better than average, and included a great benefit package, including profit sharing and health care. The best part was that there was no union involvement. Concurrent with the employment process, Keith contacted some of the people he knew and trusted in the past and offered them supervisory positions. The key slots were filled with known quantities, so Keith felt confident that the working relationships would be good.

Beth and Cindy had made a couple of trips to Atlanta in the leased business jet the company now had, and helped to set up the training program for each of the stations in the manufacturing process. There were twelve different packages in all, and Keith leased space in hotel conference rooms to conduct the training. He used the credit card for everything, and one day decided to see how much money he had spent. The total was over two million dollars, and was staggering to him, but didn't even draw a second glance from the company.

The factory was completed ahead of schedule, and installation of the robotic equipment went along concurrently with the final building phase. Immediately after the last county inspection, at which they issued the certificate of occupancy, Keith brought in the supervisory personnel, and the key line workers for a full up demonstration of the equipment. Eddie and Bill were both there, and though both knew more than Keith about the equipment, they allowed him to conduct the equipment tests.

He had done a good job of supervising the installation, and he passed the baton to the lower ranking people to explain their part in the overall manufacturing process. It was a good example of true leadership, and it was obvious that all the workers thought a lot of Keith.

"When do we expect to be up and running for real?" Eddie asked.

"I want to have the entire work force in here for a demonstration before we start production. Probably in three days. The raw material pipeline is in full swing, and I have leased a couple of tractor trailers to get the first shipment to you. I expect to have that on the road by the end of the first week."

"Everything is ready in Albuquerque, so when we get them we will be off and running. I am starting the publicity campaign the first of next week. I have the major networks set to film a demonstration of the prototype. I am going to use one of them as the test subject, so there's no question of this being just a gimmick. The big boys have all been aware of the activity of our company over the past couple of years, so it won't come as a surprise to them that we have a new product, just that the product is so revolutionary," Eddie said.

"I didn't ask, but what are these units going to sell for?" Keith asked.

"We did a cost analysis, and the figure we arrived at was between four and five thousand dollars. We make a lot of money and our people are well paid. The cost is not prohibitive for the little guy, and the well healed won't complain either."

"I figured it would be a lot higher than that for what people will be getting," Keith said.

"We didn't want to price the poor folks out of the market. They deserve the benefits of technology as much as the rich and we will still surpass the big software companies in earnings in the first year. These are going to sell quicker than we can produce them."

The following Monday their parking lot was filled with television trucks with satellite antennae bristling. Eddie and his partners had cleared a section of the warehouse and rigged enough electrical current for all of them to operate their equipment.

Beth and Cindy had supervised the set up of a food table and hired a caterer to manage the food and refreshments. All the networks were represented, and they had allowed the local independent station to witness the procedure as well.

Eddie showed them where the equipment was for the demonstration, and a set of lights from one of the networks was set up. All the cameras were arrayed in a single location and as soon as everything was set up, Eddie briefed them about what was to take place.

"Okay folks, let me have your attention," Eddie said. After everyone had quieted down he continued. "I have asked you here today to witness a pretty momentous occasion. My friends and I have developed a computer that is so far advanced beyond what the world is now using, that you really do have to see it to believe it. We will demonstrate the system to you today and following that I will take questions and give you more details. In order to assure you that this is not some sort of gimmick, I want you to select one from your ranks to be the test subject. I would like you to film the selection process as well. I suggest that the fairest method is to draw lots. There is a jug on the table by the computer with poker chips in it. All are white except one. Whoever draws the black chip gets the honor."

As soon as the cameras were rolling the reporters did their voice descriptions of the proceedings. The draw was conducted and the black   chip was drawn by one of the network reporters. Eddie escorted him to the table and placed the headset on him. For benefit of the cameras he explained the way the computer operated.

"The computer is based on a free flow of electrons for two way communications between the brain and the computer. The mode of interface is the human thought process. All you need do is look at the computer screen and think what you want to see, or what you want information about. As your thoughts change, so does the computer screen. If everyone is ready, I will energize the computer."

Eddie turned the power on and nothing happened for about five seconds. The screen then came alive as the network representative sat with the headpiece on. The images flashed as he tried to digest what he was experiencing. The pattern settled down as the man organized his thoughts better. Eddie allowed the process to continue for about three minutes, then de-energized the machine.

"I want you to explain to the others what you experienced," he told the test subject.

The reporter was almost speechless. "This is the most amazing thing I have ever experienced. As I thought about different things, the computer had the image before I could consciously register the thought in my own head. The questions I posed were answered before the thought even finished, or it seemed that way. This is so far beyond anything we now have that I can't even think of a good comparison."

Eddie took over the explanation. "The electrodes on the headset interact much as your ears. When you hear a sound with the ear, the brain interprets the meaning of the sound. Think of the head piece as the ears of the system. The brain puts out a signal in the form of electrons. The headpiece picks up the signal and transmits it to the computer, which receives the signal and interprets it. The result is sent to the computer screen, and at the same time back to the headpiece for the brain to interpret. All this is done seamlessly at the speed of light, and the machine is only limited by the ability of the person wearing the headpiece to absorb the information. Now, questions please!"

One of the other network newsmen asked the first question. "I don't see any interface for an internet connection. How did you get the responses without that?"

"This computer does not require an internet connection. It is programmed internally with all the data necessary for operation. There might be some obscure subject that it cannot respond to, but these will be very few."

"If what you are saying is true, then all the companies out there who provide internet services will go down the tubes very quickly," he observed.

"That's true, but it won't happen overnight. They will have adequate time to diversify their holdings and move into other sectors of industry," Eddie replied.

"When will these be on the market and how much will they cost?" asked another of the reporters.

"We plan to have the first units available for sale in approximately thirty days. The cost will be between four and five thousand dollars."

"Who is going to produce them for you?" asked another.

"We will produce them ourselves. My company has everything in place for production, but they will not be mass produced in the true sense of the word. Each unit will have to be programmed individually, and the process cannot be automated. The production rate will be in the neighborhood of one thousand units per week. The machines will not be sold by anyone other than my company, and orders will be taken on a first come, first served basis."

"How can people order them?" they wanted to know.

"We will publicize that information when we are ready to commence sales. One other thing that you should know is that the entire system is proprietary, and no person is authorized to tamper with it in any way. If they attempt to do so, the unit will become inoperable, and no refunds will be given. Any unit that requires service must be shipped back to us. There will not be any charge for repairing any machine, unless there is evidence of tampering."

"What exactly do you mean when you say inoperable?" the local independent reporter asked.

"I mean that the unit will look the same, but will not work. Everything in the system will stop working. It will not accept power any longer, and all data will have been erased."

"Isn't that pretty harsh?" asked another reporter.

"I don't think so. We are simply protecting our investment. If people cannot agree to those stipulations, then they cannot purchase one of the computers."

"Then you are worried about someone reverse engineering the computer?" the same reporter asked.

"Of course we want to discourage any attempt to do so, but we are not worried. Let me show you why," Eddie said as he removed the cover from the computer housing." He motioned the cameras to get a close up of the inside of the machine."

"What do you see that can be reverse engineered? There is nothing there but wires, and the interface of the wires to mate with the headpiece and printers. I defy any engineer to duplicate this set up in a manner that will work with any of my other components."

The reporter who had hooked up for the test said, "People are going to be clamoring for these, and if you limit your production you will create a scalpers market for reselling them."

"It isn't a matter of limiting production; it is a matter of how quickly we can produce them. Right now, we are still in the estimating stage. We don't know how many we will be able to produce in a given amount of time. It could go much faster than we predict," Eddie told them.

"How are you going to handle the orders?" one of the network reporters asked.

"We will set up a web site strictly for that purpose. It will be designed to take a certain number of orders per day, and the requests will be filled in the order in which they are received. All the procedures will be spelled out to the media well in advance."

The networks had intended to tape the demonstration for viewing later in the day. However, the producers were watching a remote feed, and when Eddie had explained about the system for choosing the test subject all had chosen to go with a live feed. The entire proceedings had been broadcast live. The phones at every network were ringing off the hooks by the time the demonstration was over. Most wanted additional information about the machines and how to get one, but some wanted to know where the demonstration had taken place.

The television producers were already on the line to their reporters directing them to set up follow-up interviews. Internet blogs were immediately inundated with news of the new computer, and various agencies of the government were convening meetings to discuss the implications of the new development.

The public at large had not been this excited over anything since the Y2K scare, when people thought all computers would crash when the new century turned over.

Eddie and his friends had anticipated a hubbub, and had hired private security for the property. Now that they had let the cat out of the bag, they were going to be besieged by reporters, government agencies, and other companies wanting a piece of the action.

Fortunately, the new building was completed and they were scheduled to move in at any time. That facility provided much better security, and it was easier to control access. The entire property was enclosed inside a ten foot high chain link fence with razor wire atop it, and the gates were reinforced steel, with a substantial guard house and security guards controlling access and patrolling inside the perimeter. It might seem like a bit of overkill, but they wanted it to be obvious that they took security very seriously.

After all the reporters had left, Bill suggested that they all sit down and review their plans. It was already obvious that a thousand units a week was not going to be nearly enough to meet demand. If they could produce a million a week, they would be backlogged for two years, at least.

Beth started the discussion. "There's just no way we four are going to be able to program fast enough to even begin to fill the demand."

"I am really reluctant to bring anyone else in on that part of it. It is the one thing that could compromise the project," Eddie said.

"I'm wondering if we should even be concerned about that. You said yourself that there are enough traps and blind alleys in the design that it would be extremely unlikely if anyone could duplicate the system. As for someone stealing a copy of the program, isn't there some way you could design the programming disk to self destruct after a specific amount of time?" Bill asked.

Eddie responded, "Yeah, I could do that. If I copy the disk at eight and program it to remain active until five, or something like that, it should work. We can also screen potential programmers more closely and set up a camera system to monitor their work."

Beth said, "I think we should seriously consider doing that. The four of us could then oversee the operation and have some freedom to supervise the sales and shipping aspects."

Cindy added, "Let's get moved to the new facility and then get serious about setting up the production line."

"We also need to set up the procedures for taking orders. Is the web site set up to hand off to other processors so we can take the orders quickly?" Eddie asked.

"I think that's pretty well on track. We have over fifty processors, and they are really fast. All the person placing the order has to do is enter their zip code and the program enters everything else and just asks for verification. The credit card number is back checked against the address before approval, so all the checks and balances are in place," Cindy said.

"Okay, we move tomorrow. A couple of U-Haul trucks should be enough to move what we want to take with us," Eddie said.

Chapter 10

Clarence Woodman had been watching television when the networks all switched to live coverage of a computer demonstration in New Mexico. He turned the sound up and listened as Eddie Casteen explained the theory of operation of the computer and as the reporters drew chips for the privilege of being the test subject.

It was all quite dramatic and Woodman wondered what was so special about a computer. When the demonstration started, he knew that the four were the ones he had been told to look out for. The theory of the system was beyond anything anyone without special abilities could comprehend.

Now that he knew for sure, he would have to tell the alien. He was dreading that chore, because he felt sure that he would be ordered to kill them, and he had no stomach for that.

He got wrapped up in the demonstration and follow-up questions and before he knew it an hour had passed. Eddie had been forthcoming with his explanations and anyone with half a brain could see the implication of the invention on information technology. Clarence was about to take a back seat to someone who would outstrip his earnings in less than a year. This was definitely what he had been told to look for.

He called the people in New Mexico. "You can have your team stand down now. What I feared has happened, and there's nothing we can do at this point. I may have some other work for you in the near term, so keep in touch."

He thought about his situation and knew that he would have to contact the alien again soon. He would wait for the debut of the new computer to see exactly what they had come up with. Try as he might, he couldn't see where this was all going.

If the four had been contacted by one of the aliens, then it stood to reason that they had been given the same powers he had. That was evident in the things they were doing. The drug they had developed for the cure of AIDS could have been a result achievable by someone without super smarts, but the computer he had seen demonstrated on television was way beyond the capabilities of the average human being.

He didn't think he had anything to be worried about from the physical standpoint. These were simply kids who had super powers and were using the brain power in a good way. Nothing they had done since he had been having them watched indicated that they had any evil intent.

Based on his own background, he believed he could tell enough about that aspect of life to know if they had been into anything shady.

He considered trying to contact them to see what he could learn in a face to face meeting. He didn't even know if they knew about his existence or not. If they did, they had not made any effort to do anything either to, or about him.

The decision to contact the alien was a no brainer. He just had to decide when he was going to do it.

It could be put off for a while, so he decided that he would not worry about it now.

Chapter 11

The four had made the move to the new building the following day. It had only taken them a couple of hours to gather the things they would take with them. The computers and tools, along with file cabinets were the main items. The computer hardware that had arrived for storage would be moved by others in larger trucks.

They had purchased new office furniture for the new offices and all the utilities were up and running. They had been watching the monitor of the watchers since installing the camera, and now, for the first time, they noted the empty office. None of them could envision the surveillance continuing once they got into the new building, but it had ended before the move. That may have been bowing to the inevitable, but they didn't think so.

The more likely conclusion was that the announcement of the new computer had been noticed by Woodman, and he had decided to stop the surveillance.

Beth and Cindy had drafted an advertisement to place in various papers around the country, searching for candidates to do the programming. They established pretty high educational standards and stressed the absolute need for a clean background and the requirement to pass a polygraph and background check.

They only listed a web site they had designed for responses, and anyone showing interest was run through the police records before any follow-on action was taken. Only one or two from each area was chosen. They wanted a total of twelve and it would not take long to do thorough checks on them through the agency they had hired. The salary was not listed in the advertisements, but once they were contacted they were offered very attractive benefits packages and free housing if they were willing to make the move. It was not at all hard to find twelve people who were willing, even not knowing what the work would entail.

It took less than two weeks to get everything in place. While the others worked behind the scenes, Eddie was swamped with requests for interviews. Everyone was anxious for the debut of the innovative computer.

The web site was set up, though not activated, and the handoff procedure was checked to make sure it worked as it was supposed to. The amounts of money to be processed was more than any one bank could handle, so they used ten different banks, with ten of the processing sites linked to each one.

The people they hired to do the programming were brought in and taught how to process the units. As they came off the assembly line they would be moved by hand to one of ten conveyer belts in sequence. This would give the programmers the time to program the unit before the next on his belt arrived. The process only took a matter of seconds. Eddie had bought fifty of the mannequin heads used by department stores to display hats. Into each he had inserted a processor with wiring to match up to the different contacts of the headpiece.

The programmers simply placed the headpiece over the mannequin head and inserted the disk. The loading was accomplished in less than two seconds. The completed units were then taken by another conveyer belt to the loading area where they were packaged and stacked.

It appeared that the process was going to work as planned, and Eddie did a count to estimate the number of units they could complete in a day. The number was close to three thousand. That was 300 million dollars a day!

They announced the date that they would start to manufacture and identified the web site for people to place orders at the same time. The hosting site was at maximum capacity within minutes. The system was designed to produce the shipping label when the order was taken. They tried to streamline the process as seamlessly as possible, but there were changes that had to be made that slowed them down. They only produced two thousand finished units the first day of operation.

They did not operate around the clock, but only during the regular working hours. The down time in the evenings and at night they used to service equipment and make the changes to the physical layout that they needed. One of the things that had to be improved was the final step after programming. To get from the programmers to the packers and shippers, the capacity was not there to keep up.

The problem was solved by installing another conveyer belt and hiring more people to do the handling. So far, the other plants were keeping pace, and there was a constant stream of semis in and out of the facility.

Because of the disparity in locations of the buyers, they decided to go with a fleet of smaller trucks for the deliveries. Eddie contacted the auto manufacturers and ordered a hundred vehicles from each of the three biggies. He asked Beth and Cindy to line up drivers. It was not a very cost effective way to operate, but with 300 million a day in sales, it wouldn't hurt them too much.

After the second week, they started using the commercial delivery companies as well as their own fleet.

The product had been an instant success. No one who bought one of the units would ever want to go back to the digital computers. Testimonials didn't need to be solicited. The news was full of stories about how the speed of the machines had impacted different businesses.

After four weeks and almost nine billion dollars in sales, the process was running smoothly. Eddie produced the programming disks each morning when the crew came in and collected them in the evening when they left. He immediately destroyed them, after checking to make sure the disks were the proper items. It had not escaped him that someone might try to switch disks on him, and he was very diligent in that task.

The orders were still coming in as fast as the system could process them and showed no signs of slowing. They had not had a single malfunction in all the machines that had been built, which was amazing even to Eddie. Human error would have to rear its head sooner or later.

They were so busy that Clarence Woodman had slipped their minds, but as they settled into a routine Cindy mentioned him and the fact that he was going to have to be dealt with.

The four switched off with the programming staff, but one of them had to be there all the time. It was decided that Bill or Eddie would contact Woodman and ask him to come for a visit. He surely suspected that they knew about him, and the confirmation would not affect the situation one way or the other.

By the end of the second month the backlog of orders had been filled and the system was keeping up with new orders. The employees were happy because Eddie had decided to pay them monthly bonuses at both locations. The bonuses were larger than their paychecks, which were generous of themselves.

They still had encountered only minor problems with any of the units they sold. The problems they did have were mostly human caused, so the reliability was excellent.

They were sitting around the office after hours one evening talking about how to approach Woodman. Beth said, "Why not just call him on his cell phone and invite him out for a visit?"

Getting the number was easy enough so that is what they did. The call was on the speaker and when Woodman answered Eddie said, "Good evening Mr. Woodman. This is Eddie Casteen. I believe you know who I am."

"Yes indeed. I have been watching your business endeavors. You aren't calling to offer me a piece of the action are you?"

Eddie laughed. "Not hardly, however there are some things we need to talk about. How would you feel about paying us a visit out in sunny New Mexico?"

"What kind of things do we need to talk about?" he asked cautiously.

"Well the people you had watching us for two years would be an indication that you want something from us. We have a pretty good idea that it might concern space age technology, and we would like to compare notes."

"What's to keep you from just getting rid of me?" asked Woodman.

"If that was our objective you would have been gone a long time ago. You should know from your youth that it isn't that hard to make someone disappear, even someone of your stature. We simply want to talk. I think the conversation will be enlightening to both us and you. There are a lot of unknowns about both our situations that need to be resolved."

"When do you want to meet?"

"Whenever you choose, as long as it is after five in the evening. We operate on an eight hour day, and close down production at five."

"You have the hottest product in the universe and you only operate eight hours a day?" Woodman queried.

"You got it. Nobody is going to undercut us, so why the big rush?"

"You have a point there. How about tomorrow night? I will fly in during the afternoon and call you before I show up at your front door."

"We look forward to it," Eddie said.

Chapter 12

Clarence Woodman arrived in the later afternoon the following day. He had a private business jet, and called Eddie when he touched down in Albuquerque.

"I suppose you can find our office. I will have security bring you to us when you arrive. We are going to order some food and have dinner here if that suits you. Is there anything in particular you would like?"

"Whatever you choose will be fine. I will see you in about half an hour."

When Woodman arrived the introductions were made and the group moved to the conference room where the food had been laid out on the table. As they loaded plates and started to eat, Woodman asked, "How long have you known about the people I had keeping an eye on you?"

"Almost from the start. The ploy to snatch Beth was not smart. That alerted us to the fact that something was not right, and the circumstances of the abduction coupled with the time element gave us an idea that whoever did it was close by. It was just a matter of watching our mirrors after that. I suppose you should know that we had video coverage of your people for the last eighteen months, and knew their every move."

"I should have figured that out when you didn't squawk about the abduction incident."

"I don't know how to delicately get into the subject we need to talk about, so I will just be blunt. We four had contact with an alien, and we believe you have had a similar experience. The things we were told do not seem to make a lot of sense to us, and we wonder if you would be willing to exchange stories of your encounter so we can try to better understand what has happened?" Eddie asked.

"Let me ask you a question first," Woodman said. "Did you have some gizmo similar to the headpiece on your new computer system placed on your head?"

"Yes, we all did," Beth answered for them.

"And did you receive specific instructions about anything?"

"No just that we had been chosen to save the world from total destruction at some future time. The entire encounter didn't make a lot of sense to us, and we have been puzzling over it since the encounter. We even went back and asked for some clarification, but didn't get a lot of help. The thing that puzzled us most was the alien's premise that we would be operating against another alien who had the same capabilities of our benefactor, but who supposedly had evil designs on the planet. That didn't make sense to us because anyone with any sense at all would not work toward destruction of the planet he lives on. Did your encounter involve any specific instructions about what you were to do?"

"My only instructions were to look for someone like you four and report when you showed up. I have been doing that for forty years, with a visit to the alien yearly. He or it, has not asked that I do anything else, and frankly, I am a bit afraid of what he will ask of me when I tell him you have shown up."

"You think maybe he might want to get rid of us?" Eddie asked.

"That thought had crossed my mind. I don't have the foggiest idea about what we are involved with, but at the time I was contacted, I was destitute, and probably headed for a life of crime. I jumped at the chance for the money without any thought to what the consequences would be."

"We were told that someone like us might already be in place, or would be at any time, but we were not given any instructions about how to deal with it. He only told us that you would be the tool of another alien trying to stop us from saving the world. If I was writing a story about this, the dialogue would have seemed laughable."

"Let me ask you another question," Woodman said. "When did you get the idea about the computer you are building now?"

"I don't know," Eddie said. "After the encounter, when we discovered how much smarter we were, all kinds of ideas crossed my mind. This one seemed to stay with me longer, and was more persistent."

"I know what you mean. Even a dummy like me became a genius in the blink of an eye, so the four of you working together are unlimited in your abilities. The AIDS cure Bill formulated and the way he put it on the market went a long way toward convincing me that you were not adversaries in the sense of evil intent. He could have made a mint on the drug, but chose to make it affordable for those who really need it. The same is true of the computer you are producing now. You could have held the concept close to the vest and gotten hundreds of thousands of dollars per unit if you chose to do so, yet you priced it so that the less fortunate could benefit as well."

"Well, we are still making an awful lot of money off it," Cindy said.

"But the point I am trying to make is that I don't see any ulterior motives in anything you guys have done, and I wonder how the alien will react when I tell him about you."

Eddie said, "I don't think either of us is reacting the way the aliens thought we would. I don't imagine they envisioned this meeting for example. Our dilemma from both sides is that we don't know the purpose or motive of either of our situations. And further, we don't know the consequences of any actions we will take or have taken."

"The aliens seem to be treating the whole thing like a game. I am not so sure I believe what we were told about them not being able to interfere with us. It stands to reason that if they contacted us once, then they could do so again at will. We were told that we had to be in a specific location to make contact, and maybe there is some truth to that, but I would bet that if they can arrange contact in one place, then they can certainly arrange it anywhere they want," Beth said.

Bill said, "The thing we know for certain, is that they are advanced so far beyond us that we cannot even comprehend the possibilities of their capabilities. We don't know anything at all about their physical makeup, or even if they exist physically. The one who contacted us was sort of opaque, without discernible features, so they could be composed of some elements we don't know about."

"The thing is, we have been concentrating on the end of the world scenario Joe told us about, and the opposition we would be facing, that we haven't given a lot of thought to other aspects of the meeting," Eddie said.

"What are you talking about?" Clarence asked.

"Well, the remoteness of the location for one thing. It doesn't make sense that a being as advanced as they obviously are would choose such a location. He could just as easily have beamed someone else up from his bed, or a drive-in movie, or anyplace a lot more accessible. Instead, he chose to contact us from a place that humans would hardly even pass by accident. I imagine your encounter was in a remote location as well," Eddie said to Clarence.

"It was not as remote as yours obviously was, but it was not a well traveled location either."

Cindy said, "I believe the only way we are going to get any answers is from the aliens. Either we need to go to our location and take Clarence along, or he needs to go to his location and take us along. We need to resolve this before we can even begin to understand what it is all about."

"If their aim is to advance human knowledge and technology, then I can see where you guys come into the picture, but in my case I haven't done anything to justify their actions with relation to benefits to society. One or the other of them is not being forthcoming with us," Clarence said.

"Would you be willing to broach the subject of a meeting with us when you contact the alien?" Eddie asked.

"I think I have to. He has told me that my sole function is to identify you when you appeared and tell him. I will at least get some idea of his intentions at that point."

"I don't think we should do anything on our end of things until Clarence has his meeting," Bill said.

"I agree," Eddie said. "I have an apprehensive feeling about this whole thing now. I am beginning to wonder if the idea for the computer was mine, or if it might have been implanted by the alien. Even with our expanded intelligence, we are so far behind the aliens that it would be laughable, if the consequences were not so serious. One thing is for certain though, the end of the world line is looking more and more like a hoax. I can't come up with a reason for his telling us that, but I sure can't see anything that would cause that result either."

"Maybe it was just a ploy to get us moving in the direction he wanted us to go," Beth said.

"In a way I wish we had never had that encounter," Cindy said.

"What's done is done, and we have to find a way to deal with it," Eddie said. "You can't knock the money end of it though."

Clarence said, "I will fly back east and meet him tomorrow. I will call and let you know the results of the meeting."

"Find out if he can meet with all of us, or if it will have to be only one. I am not so sure our alien was truthful about that either," Beth said.

"Now that we have met, I am sorry we didn't get together sooner," Clarence said.

"Things would certainly have had a different outcome," Bill told him.

"I keep having a feeling that there is something we are missing, and that whatever it is might come back to bite us in a big way," Eddie said.

Cindy said, "I know what you mean. It's like a little tickle in the brain that won't go away. It's a combination of apprehension and guilt. The apprehension is because we don't know what is coming, and the guilt is because we should have seen it sooner, whatever **it** is."

"Well, maybe my meeting will shed some light on the situation. I will be in touch," Clarence said as he got up to leave.

Chapter 13

The next day was Friday, and they would have the weekend free if they needed to all go back east. They had leased a business jet, so transportation would not be a problem. After Woodman had left the previous evening, the four had a frank discussion about their situation. All seemed to feel that they were being used in some way, but for all their intelligence, could not figure out how, or why.

Beth said, "I haven't felt this helpless since we had the encounter with Joe. Even the time we were being watched and followed everywhere we went didn't bother me much. But, something about the uncertainty simply unnerves me."

"I think it is affecting us all the same way. It is the feeling of being used under false pretenses that rankles all of us I believe," Bill added.

"It's more than that," Eddie said. "I don't believe the apparent randomness of our encounter with Joe was as he stated. I think we were chosen for some specific reason. Probably one factor was our relative youth. We would not be as worldly, and therefore less likely to look for a motive, other than what we were told. Add to that the awesomeness of the added intelligence and we were like kids in a candy store. The key to this whole thing is motive. Either Joe, or Clarence's alien are lying through their teeth, if they have any. It will be really interesting to hear from Clarence after he contacts the alien again."

Cindy said, "If he agrees to meet with only one of us, I suggest that Bill be the one to go. On the other hand, Joe might have been telling the truth when he said that the other alien could only meet with a single person. If that's the case, then we will have to use Woodman as a go between. I got the feeling that he is not a bad guy, and that he was sincere with us."

"I did too," Beth said.

"Regardless, we have pretty much negated any action he could take against us by laying our cards on the table. I think he was really worried that his alien might have wanted him to take care of us in a permanent way," Eddie said.

"You mean kill us?" Beth asked.

"Yes, and that is still a possibility, though I think remote," Eddie replied. "Even if he tells Woodman to take care of us, I don't believe he will do it, or even hire it done."

"I agree," Cindy said. "We have too much security in place, and we can never go anywhere without the press hounding us, so it would not be easy, even if he consents to make the attempt."

"I think we are all agreed that he is not a threat," Bill said.

"I guess we wait to hear from Clarence then," Eddie said.

They called it a night and went home.

The following day was like the previous two months had been, except all their minds were elsewhere as they continued with production of the computers.

The total sales now topped three billion dollars, most of which was profit. The production of the components at both plants was less than one hundred dollars per unit. Even factoring salaries and bonuses, plus the cost of the property and buildings, the profit margin was ridiculously high; something like eighty percent.

When they finally closed down the production line for the day, and secured the facility, they still had not heard from Clarence. They assumed that he would meet with the alien at the earliest opportunity. They had no idea where he had to go for the meeting, and if it was the same as their procedures for contacting Joe. If so, then it would have to be at a specific location. It was probably somewhere on the east coast, since that was his normal stomping grounds.

They went out to dinner and were just finishing the meal when Eddie's cell phone rang. It was Woodman.

"Well, I made the contact, and you won't believe this, but before I could even broach the subject of a meeting, after I told him about you, he suggested that I contact you and ask if you would be willing to meet with him. I asked him why, and he said that it was a very complicated matter, and that it would be best if he didn't try to explain until I could get you to agree to meet," Woodman related.

"Did he say he wanted to meet with all of us, or just one?" Eddie asked.

"He asked for all of you to be present. I asked him why just one wouldn't do and he said that what he had to say would affect all of you, as well as me, and that we should all be present."

"When does he want to meet?" Bill asked.

"He said the sooner the better. I got the feeling that he is expecting something to happen soon and doesn't want to waste any time."

"Okay, we will leave tonight and be there in the early morning hours. Can we have a daytime meeting, or would darkness be better?" Eddie asked.

"Probably either will do. The location is kind of remote. I don't mean that it is in the wilderness, but is far enough off the road that we won't attract attention if we make the rendezvous in the daytime."

"Where should we meet you?" Beth asked.

"Philadelphia will be the closest, and will be fairly easy to get into in a business jet. I will meet you at the private terminal in Philly. When you file your flight plan, they will have the time of arrival."

The four went back to their house, changed clothes and packed for the trip. They would not be gone more than two days, so the packing was done quickly.

Eddie called the airport and told them the plan and asked that the flight crew be alerted for take-off within the next two hours. He gave the location, so the pilot could file the flight plan when he arrived at the airport.

The entire group showed some signs of excitement mingled with a bit of fear of the unknown. They realized that they would be sailing in strange waters, and though they needed the information, were afraid of what they might find out.

The pilot had the plane ready and the flight plan filed when they got to the airport. The flight was about three hours and with the time differential, they arrived in Philadelphia at five o'clock in the morning. Clarence Woodman was there to meet them, as he said he would be, and was driving a rented Dodge Grand Caravan, which would hold them all, as well as the luggage.

Clarence got everyone loaded up and headed north. He gave them a brief recap of his first contact some forty two years previously, telling about his life to that point, and why he jumped at the chance to make a lot of money. "I was probably a lot like you guys. I was about your age when it happened, and my life was not very pleasant to that point. I was running from some organized crime enforcers trying to collect money I had borrowed, and was hitch hiking to New York to try to make a new start. There was not much traffic on the road, and when I spotted a barn that looked like it might provide some shelter, I headed for it. That was when he made contact. I was scared out of my wits at first. When he explained what would happen, and what he wanted me to do, I didn't see any down side, unless when someone showed up he wanted me to get rid of them. I figured that was a bridge I would cross later, and the present was the important thing at that moment. Everything worked exactly like he said it would, and I have lived a life of luxury for over forty years. I am glad it didn't end the way I thought it would, both for your sakes and mine."

Eddie in turn told Clarence about their encounter and what the alien had told them about the destruction of the earth, and how they were to somehow figure out how it was to happen and stop it in some way. The story was plausible to them and they didn't doubt the story for a moment, though they could not come up with anything that led them down that path.

They drove for over an hour before Clarence turned off the highway onto a farm road. He drove only about two hundred yards off the main road and parked the van. "I don't know how he will do this, but I usually go to the location and he makes the contact. I don't think my body goes anyplace else, just my mind. We will give it a try the way I normally do it and see what happens."

He led them to an area under a grove of trees and when he got to the location they all stopped. It was only a matter of seconds before they all found themselves in the presence of the alien.

"Greetings," he said. "I have waited a long time for this meeting, and frankly had about decided that it would never occur. Clarence has been awaiting your arrival for over forty of your years. That my judgment was vindicated is pleasing to me. I suppose all of you would like an explanation?"

"We surely would," Eddie said for the group.

"First of all, I am not from your galaxy. My species exists in a parallel galaxy that is far away. You could not get there in your lifetimes, though when your kind solves the riddle of time travel it will be possible."

"That's all well and good," Eddie said, "but we would like to know what is going on, and more importantly, what part do we play in it?"

"I don't blame you for wanting the information, but first I need you to understand why my species are here, and what that means to you five in particular. You see, we do not require nourishment as you know it, and as a result can exist for a very long time. We are a combination of elements brought together in a certain way and draw our nourishment from other elements. This allows us to exist for thousands of your years, and when we become outdated, we can be regenerated. Our population is controlled so that we do not overpopulate our home planet."

"Why are you here on earth, or above earth, or wherever we are?" Beth asked.

"We send out explorers in groups to other galaxies to see what life is like on planets where we discover living creatures. We study them and send reports back to the home office, so to speak."

"What is the purpose of the exploration?" Bill asked.

"To observe mainly. We are cautioned not to interfere with the evolution of the planets we encounter, but can interface on a limited scale, such as you five. Our leaders allow us to impart knowledge to other species if it is such that it will help the species advance without calling attention to the fact that there are other beings elsewhere that you don't know about. We have done this on occasion here on your planet, and it has worked out well. The problem is that some of our explorers sort of look at this procedure as a game. I fear that this is one of those cases. One of my brethren has decided to exert too much influence, and I am afraid that if he has his way, your planet will cease to exist in its current form."

"What do you mean in its current form?" Eddie asked.

"I mean that he would like to annihilate the majority of your population just to see how long it takes the remnant to rebound to its current state."

"And I suppose you are going to tell us that he is going to do that through us," Cindy said.

"I think that is a distinct possibility," the alien countered.

"Would it surprise you if I told you that one of your kind told us the same story, except he was the good guy and you were the villain," Eddie said.

"Not in the least. That is exactly the way I would expect him to make the attempt."

"So how is he likely to do this?" Bill asked.

"He has probably implanted some idea into your minds that he hopes will achieve his ends. It is hard to say exactly what he will do, but rest assured he will make the attempt."

"I think he already has. I had an obsession about a new computer concept which operates on the theory of free flowing electrons. In retrospect, the headpiece interface I use is very similar to the gadget he used on us when we first met."

"It is very possible that he will use your computer in some manner to achieve his desires. You have already started to produce the computers, and they are probably so advanced beyond what you have used in the past that everybody will want one. As soon as they are in widespread use he could initiate whatever action he has planned."

"He told us that he could not affect anything directly. Is that truthful, or was he trying to lull us into a false sense of security?" Eddie asked.

"It isn't that we can't have a direct affect, but that the way the mission is structured does not allow us to do so. He could disregard the directive, but I don't think he will. There are very stiff consequences for doing so."

"How so? Who is to stop him?"

"If he takes such action, then the others on this mission will destroy him and his vehicle. It is hard to explain since your race is limited in understanding, but the mission profile will reveal to the others what has happened, and he knows that."

"Do you communicate with him?" Eddie asked.

"We are supposed to act independently. We are sort of evaluated on our actions after the mission is complete, so basically we compete to see how effective each of us was in helping to advance the target civilization."

"So it's like a game. When the mission is complete you tally up your points for progress and the winner is the one with the most points?" Cindy said.

"That is not entirely accurate, but it is close," the alien told them.

"How do you determine if he has interfered?" Eddie asked.

"If you will consent to a mind probe, I might be able to see what he implanted into you that will allow the action that he wants to occur."

"How will you do that?"

"With the same device he used on you the first time."

"And that might tell you what his intentions are?"

"It is possible, but not a sure thing."

"Then let's do it," Eddie said.

The alien placed the headpiece over Eddie's head. It only took a few seconds, and as before, he had not felt anything."

"I can see that he did in fact implant the idea for the computer. The particulars he left you to figure out on your own, but I did detect some distinct characteristics he wanted you to use. I hate to say it, but if you used the information, then the computers will become deadly to the users at some point in time. I cannot tell exactly what the time element is, but when the computers arrive at a given level of use, the system will cause the user to go mad, or will reduce them to a vegetative state."

"What was the specific thing that causes this?" Eddie asked.

"It has to do with the way you built the headpiece. If the wiring is constructed in a certain way, then the system will self destruct at some point, and the brain of the user will also self destruct."

"Can you explain to me the particular procedure that would cause this?"

"I can implant it in your mind and you will be able to understand it."

"Then do so. I can't let that happen. The hard part now is going to be convincing the ones who have them that they have the potential to be lethal."

The alien placed the device over Eddie's head again for a short time. When he removed it Eddie was lost in thought for a few seconds.

"Just that small a thing could cause all the damage you talked about?"

"Yes. You realize that the design of your computer was based on the device we use to boost your intelligence. It is not exact, but the principles of this device is what you designed into your computer."

"I knew from the start that there were a lot of similarities, but I didn't connect them with some ulterior motive. I can change the design to eliminate what you showed me, but that doesn't solve the problem of getting the ones now in use off the market."

Beth chimed in, "You could say the machines already sold have a defect and that they will be replaced at no cost. I don't think we will be able to get all of them back, but we should get the majority."

Clarence had been quiet during the exchange, but now spoke up. "Am I to understand that the computers you have sold so far are potentially lethal?"

Eddie answered, "That's it in a nutshell. There are almost ten thousand of them out there already and it will take us a couple of months to produce enough to replace them, not counting the time it will take me to redesign the headpiece."

"The good thing about the situation, if there is such a thing, is that we have records of everyone who purchased them and we should be able to locate them all," Cindy said.

"What exactly is the thing that will cause them to be dangerous?" Bill asked.

"It's in the wiring for the headpiece. One simple wire that shouldn't be there could lead to the destruction of a large portion of the human race. I don't understand exactly how it will do what our friend here said, but I know exactly where the wire is located in the design. I am not convinced that the single wire is the entire problem, and before we revise the headpiece and go back to production, I want to look at it more closely."

"I have a question," Beth said. "If we meet with our contact again will he be able to tell that we have been in contact with you?"

"No, he will not," replied the alien.

"Then I suggest we go back to him and try to probe a bit more about his destruction of the earth scenario. He knows we are obsessed with this, and I don't believe he will think it strange that we are seeking additional guidance."

"Do you think we should tell him we have identified Clarence as one of us, or would that make him suspicious?" Bill asked.

"I think he would be suspicious if you neglected to tell him that."

"Another question," Beth said, "If we come back here will we be able to contact you?"

"Yes, of course. When your contact told you that I could only interface with one individual at a time he was trying to lull you into a false sense of security. He intuited that I would only contact a single individual to keep an eye out for the people he interfaced with. Thereby he advanced his standing with you four when you learned about Clarence."

Eddie said, "I have to ask - could it be possible that he has interfaced with others to have someone to keep tabs on us? I ask because at the beginning, we had no money to do anything, and he arranged to have Bill come into an inheritance from an aunt he hardly knew. Bill called to verify with the lawyer and was told that everything was on the up and up. He had to have someone in the office to do the verification."

"It is very possible. He could have been in contact with others for a number of years, just as I have been with Clarence. None of us are limited in the number of people we can interface with, so it is probable that he had someone waiting in the wings for just such a circumstance."

"I have another question," Cindy said. "What's with the specific location we need to be at to make contact? I would think your kind could choose to make contact anywhere you wished."

"That is true, but because we are physically located many light years away from your planet, we must isolate a specific location to project to. Think of a beam of light, or a laser beam. The strongest point is the area where the beam is focused. With us it is much the same. We project along the trajectory of the beam, and the point at which we interface is the point at which the beam has the best focus. That is the reason that most of our contact with your world is from desolate locations. It simply takes too much energy to move outward from the beams focal point."

"Okay," Cindy continued, "then you are not all knowing and all seeing. I mean, you must have some method of knowing what goes on here, but you are not omniscient. You can't be aware of everything that goes on, but can you zoom in on specific things?"

"Only in a very general way. We can observe from a distance, and extrapolate the advancements made by your species based on these observations. We cannot focus on an individual to follow their movements for example, but we can observe the types of transportation, infrastructure, and so forth that give us indications of your advancement over a period of time. If you are worried that your individual actions can be observed, then you have no cause for concern."

"Am I to understand that unless we have the headsets on, there is no way you can read our thoughts?" Eddie asked.

"That is correct. We can read much from your expressions, so be careful to keep a straight face, as your saying goes, when you interface with him."

"Then, I guess we should try to deal with the problem of the computers already out there and visit Joe again," Eddie said.

"Is there anything else we need to be concerned about with Joe?" Bill asked. "And on that subject, what should we call you?"

"You can call me Tom," the alien said.

"Should we set up some schedule to visit you, or just show up when we need to talk?" Cindy asked.

"Whenever you need to talk, I will always be here."

"Well, we thank you for agreeing to meet with us, and for the information you provided. There might still be widespread loss of life, but at least we will be able to stop the wholesale destruction of our race," Eddie said.

Clarence said, "Whatever you do to try to stop this guy, I want to be involved."

Beth said, "Thank you Clarence. We will keep in touch and let you know."

The group headed back to the Philadelphia airport and parted company with Clarence. Eddie was so preoccupied with the computer problem that he hardly said a word all the way back to Albuquerque.

Chapter 14

Once the four got back to their home they discussed the problem at greater length. They all knew that it would be impossible to pull the computer off the market. It had already proven to be the greatest advancement of all time in information technology and the public outcry would be so great that they could not even predict the outcome if they attempted to do so.

Eddie said, "I know from the contact with Tom what the major problem is, and it will not take much to fix it, but I am still worried that Joe might have steered me in other areas of the design. If I make the simple fix and we go back into production, and there is something else awry we will be back to step one. I just don't know how to approach the problem to ensure that doesn't happen."

Bill asked, "Is there any way the rest of us can help?"

"I don't know. The computer itself is nothing but a medium to translate what is coming from the headpiece to a visual mode. I don't see how anything in that design could cause any lasting effect to the user. Can any of you envision the headpiece design based on the interface we have had so far?" Eddie asked.

All shook their heads negative.

"Then I guess I am the one who is going to have to solve the problem. It still seems too simplistic that the one wire could make all the difference in the capability of the headpiece."

Beth said, "When Tom placed the headpiece on you to get the design he surely would recognize anything else that was different, don't you think?"

"I hadn't thought about that, but it stands to reason that he would be as familiar with their design as Joe is. If something else could cause a problem he should have been able to detect it. I think the best thing to do is change the design to take out the wire he identified and get the others off the market," Eddie said.

"You don't think the frequencies could make a difference, do you?" Cindy asked.

"I wouldn't think so. Remember what a hard time I had finding the frequencies that would work? I don't think other discrete frequencies are available, so that shouldn't be a problem."

"One aspect of this problem we haven't considered, and maybe a very important aspect, is whether Joe has someone else keeping an eye on us. If we start to recall the computers and there is someone monitoring our efforts, then they could tell Joe and he might accelerate the time element for whatever he is going to do. If they can only interface with humans at one location, then we will have to have someone watch the location where Joe contacted us, just to be on the safe side," Bill said.

"That's a good point. I think we want to go back and talk to Tom again before we do anything. Let's make a list of questions. Only include the things that we think might have a direct impact on our actions. For example, can they change the location from which they interface and then go back to the original location?" Eddie asked.

"Another thing we might need to know is the distance at which they can detect our presence from the specific location at which we interface," Beth added.

"You are thinking about staking the location out and remaining at a distance at which the watcher cannot be detected," Cindy said.

"Exactly," Beth replied. "I don't see us doing this, but hiring someone reliable for the job. It might even be something we want Clarence to handle."

"Well, he wanted to be involved, and that is a way he can do something worthwhile. Somehow I don't see him as the outdoor type. Add that to the list of questions," Eddie said.

"How do you see this playing out if we manage to get the computer problem solved and get the bad machines back?" asked Bill.

"I don't have the foggiest idea. That's another question we might want to ask Tom. If we can nip the problem in the bud, how will that affect the relationship between the aliens, just knowing what Joe attempted to do?" Eddie asked.

"I hate to make another trip back east so soon, but all of these are important questions," Beth said.

"Let's call Clarence and make arrangements to go back in the morning," Bill said.

"We need to make sure our list is complete before we go back. At least as complete as we can make it based on the knowledge we have now," Cindy added.

"I believe the best way to watch Joe's location will be from a motor home, if we can get one close enough. That is some pretty rugged country and I don't know if we could get anything like that in there," Eddie said.

"If you remember the topography of the location, you might recall that there is a good sized hill not too far from where we camped in the meadow. If someone set up camp in the trees they would not be visible to anyone at that location. They would have to be outdoorsmen, but it could be done," Bill said.

"Eddie, give Clarence a call and run some of this by him. He has been dealing with Tom much longer than we have been on the scene. He might be able to give us some meaningful input," Beth said.

Eddie called Clarence. When he answered, Eddie said, "I hate to keep bothering you, but we have been hashing our problem over and have several questions for Tom and a couple for you as well. First of all, are you the outdoors type?"

"I enjoy being outdoors, but do I know anything about hunting, fishing, and hiking? The answer is no. What's on your mind?"

"We were thinking that we might want to stake out Joe's location to see if anyone else shows up to keep him informed about our actions with the computer. We won't know how this is going to play out until we confer with Tom again, but we have put together a list of things we need to know to see this through. We will fly back first thing in the morning if you want to meet us in Philly again."

"Let me know when you will be arriving and I will meet you. This is the most excitement I have had since I was being chased by the bookie's enforcer forty years ago. Somehow the consequences seem direr this time, but at least I know that I am on the right side."

"Depending on what Tom says about their ability to change locations, we will need three or four good trustworthy men to spend some time in the country, very rough country at that. Can you think of anyone who might be capable of doing that?"

"I know some people in the defense department I have worked with in the past. They are mostly retired now, and probably wouldn't be averse to taking on a chore like this, especially if the pay is good. How soon are you going to need them?"

"I want to have another session with Tom first. We can build a plan after that, based on his answers," Eddie told him.

"So I shouldn't do anything until after we get together again?"

"You can start to think about who you would use, but don't make any contact until after we get together again."

"Okay, let me know when you will arrive. We will do the same as before for transportation," Clarence said.

After he hung up Eddie sat for a few moments in thought. If someone was keeping an eye on them for Joe, then whoever it is would surely key on two flights to Philadelphia in such a short span of time. Maybe he was being paranoid, but with the stakes of the game, he figured you could not be too careful. He voiced his concerns to the others. "So what do you think?" he asked when he explained his thinking.

Beth said, "I think that is another thing we did not think of that we should have. Let's fly into Washington and rent a small plane under an assumed name to take us to Philadelphia. I am beginning to think that all our thoughts should be paranoid until we can get things sorted out. I also think we should be circumspect with our telephone conversations. If Joe has had someone in place for a long time, they will be a lot better at the job than the people Clarence hired."

"Okay, paranoia is the word for the day," Eddie replied. "And I like the idea of flying into Washington instead of Philadelphia. Even if we have to drive, it is not a great distance. We can still be back here by Monday morning. Can you guys remember the location where we talked to Tom?"

"I think he implanted the location, because I remember it very precisely. I will call Clarence again and tell him we will meet him there," Eddie said.

He made the call and conveyed the thought that someone might be listening to their conversations, so he should be circumspect when talking.

Clarence understood right away, and agreed that it was certainly a possibility. They would call when they got close and have him meet them at a designated time. Since they did not know how long it would take them to get there, they decided to fly commercial and leave at the earliest opportunity. It was impossible to get on a commercial plane without showing identification, and there would be a record of their flying to Washington, but that couldn't be helped. They could rent a plane for cash and not leave a trail from there onward, which would throw anyone off unless they expended a lot of time and legwork running down their activities after the fact.

They just had time to get to the airport to catch the next scheduled flight to Washington. They didn't even take any baggage since they planned to be back by the next night. Eddie looked up the number for an aircraft rental company in Washington before they left for the airport. He made the call while Cindy was driving them to the airport. He indicated that he wanted to rent a light plane for some aerial sightseeing and gave a phony name. The person he talked to agreed to have the plane ready at the time he specified.

When they landed in Washington, they inquired about the aviation company and got directions to their hangar. Eddie also called Clarence and gave him an approximate time that they would be at the location to contact Tom. They flew in to Philadelphia and instead of renting a car, which would tie them to the area through credit card usage; Eddie went to a used car lot and purchased an older model car for cash. He then went back to the airport and picked up the others.

When they got to the area, Clarence was already there. They parked the car out of sight of the road and the five of them again approached the grove of trees where Tom had met them.

As usual, they were greeted right away. "I suppose you must have additional questions after giving the matter some thought," Tom said.

"You got that right," Eddie replied. "We compiled a list of items we thought you might be able to help with. The first item deals with your capabilities. How far away from this spot can you detect someone?"

"As I told you, the principle of our interaction requires that we be very close to the focal point of the beam. I cannot detect anyone on the road you arrived on, and I think anything outside a one hundred yard radius would be safe, though you might want to stay a bit farther away than that."

"Okay," Eddie said, "We also had the thought that Joe might have someone watching us and reporting back to him. If I start pulling the computers off the market, someone might be watching for that to happen. If that is the case, and I think it is very likely that Joe does in fact have someone keeping tabs on us, then the minute we start the action they will report to Joe. That brings us to the question of your interactions. Can you use multiple locations, or are you tied to a single location?"

"We can use multiple locations, but not at the same time. For example, if I chose to use a different location, then this precise location would not be available to me in the future. It is simply because the devices we use for transport is not that precise. It would be next to impossible for me to return to this exact location once I have left it. The same would hold true for all of the explorers."

"Then if we set up a watch over the location where Joe contacted us we should be able to detect anyone attempting to contact him?"

"Yes."

"The next question is, how sure are you that the single wire is the only flaw in my headpiece, and is there a possibility that whatever Joe is attempting might have to do with the frequencies I used on the computers?"

"We are all familiar with the structure of the headpiece, and from what I got from the probe I did with you, that is the only unfamiliar element to me. I will probe you again if you will permit me, just to be sure the physical computer design you used cannot be subverted."

"Let's do that. I would rather be safe than sorry." Eddie submitted to the headpiece again and Tom took a bit longer than he had before.

"I see nothing else that he could use. The signals are modulated by the brain waves and are all harmonics of the basic frequency. Nothing in that part of the operation can be exploited. I only see the wire we discussed earlier as a potential problem."

"That brings us to the question about how you and your compatriots will deal with Joe. Will just the fact that he tried to do this be grounds to get rid of him, or do you have some other procedure to deal with it?" Eddie asked.

"If in fact nothing happens, then we will not be able to do anything to him until we return to our planet, and that will be far in the future. On the other hand, you might be able to do something on your own. I will not burden you with the particulars until we see if you are successful in solving the computer problem."

"I take that to mean that you will help us to get rid of him?"

"I can provide you with the tools to do the job, but you will have to reason it out on your own. I believe the first order of business is to render the current plan inoperative. Once that is done, then you will have more time to address the other issues. It is imperative that nobody else be in contact with Joe until you have the computer issue resolved."

"At least we can watch his location and prevent anyone from getting near him," Eddie said.

The others had let Eddie and Tom do all the talking. Now Beth said, "Tom, just for the sake of argument, was the computer a good idea for our species?"

"Only time will tell, but simply understanding the concept will spur a number of your more intelligent compatriots to greater technological breakthroughs, and that is progress. Joe may have managed to outshine all of us, even though his intentions were the opposite."

"Before we part company, can anyone think of anything else we need to know?" Eddie asked.

After a few moments they all shook their heads no. "Then I guess it's so long for awhile Tom."

"I wish you luck, and my portal is always open," he replied, tongue in cheek.

The five left the area and drove to a diner Cindy had noticed about five miles back toward Philadelphia. They went inside and found an isolated booth. They all ordered coffee and talked about how to set up the surveillance of Joe's site.

"First of all," Eddie said, "The location is more desolate than Tom's. It's high up in Yosemite National Park, and the only way in is by a forest road that requires a four wheel drive vehicle. The only other option would be to use a helicopter."

"I'm not much of an outdoorsman, but I can hire people who are, and are familiar with this type of activity."

"That is good, but we need one of us with them, in case it becomes necessary to eliminate someone. I don't imagine killing would be new to them, but I would hate to ask someone to take a life without knowing the reason, and we obviously can't get into that."

"If you find the people, I will go along and do the dirty work if it becomes necessary," Bill said.

Clarence said, "I will contact some people and let you know, probably by late tomorrow."

"Remember about the telephone. It might be a useless precaution, but we can't take the chance of Joe finding out that we are onto him," Eddie said.

They finished their coffee and left the diner. When Eddie and his group got back to the airport in Philadelphia, Eddie parked the car in long term parking, on the chance that they might have need of it in the future. They returned to Washington in the light plane and got on the next flight to Albuquerque.

Chapter 15

When they got back to Albuquerque, Eddie headed for the factory with Beth. Bill and Cindy returned home.

Eddie started on the redesign of the headpiece. The modification was simple, but revising the assembly line was more of a chore. He worked through the night to have the revised procedure in place for the Monday morning shift.

He had it all done before the first assembly line worker showed up. Bill and Cindy were in early and Eddie went over the changes with them. "I guess Beth and Cindy can start contacting our customers tomorrow. I want to run an extra four hour shift in the evenings starting Wednesday. That way we can continue to fill orders and replace the ones already out there with the ones we run on the extra shift. We need to devise some method for tracking the ones coming back in too."

Beth said, "I will set up a procedure for that. What are we going to do with the ones coming back in? Is there some way you can modify them and use the parts again?"

"The important thing is to get the headpieces destroyed before Joe finds out. Speaking of Joe, we are going to have to pay him a visit soon, probably before the week is out," Eddie said.

"When Clarence calls to let us know about the people to keep watch I will make arrangements to meet them and guide them to the location. I will remain on site until you guys get there," Bill said.

"You are going to have to purchase a rifle, just in case, and make sure you know how to use it," Cindy said.

"Gee, why didn't I think of that? Of course I will buy a rifle, and I already know how to use one," he replied.

The new headpiece design was explained to the assembly line workers and the explanation given was that Eddie had discovered a way to make the units more efficient. He also asked if they would be willing to work an extra four hours three days a week for double wages.

Since the company had been very generous with them, not a single person expressed reservations about the extra hours. Beth had segregated the previous sales so as to make sure all of them would be contacted with the upgrade, which is what they were calling the new design.

It was important to get all the original units back. The government had bought a thousand of them and it was not going to be easy to recover all of them without revealing the real reason for the recall. Beth was confident that she could get the majority of them back rather quickly. The problem was going to be with the ones shipped overseas or to remote locations. She would figure some way to deal with the problem.

Cindy was helping design a program to track the returns and verify the authenticity of the returned product. Each of the units had a serial number on the computer housing, but the headpieces had only the company logo. Invariably, some of the units were going to get confused, so they had to come up with a method to tell them apart.

"Do you think it would be practical to color code the new ones?" Cindy asked.

"I don't know, but we have to have some method to tell them apart," Beth said.

"Maybe we can change the logo in some simple manner that will not be noticeable to others," Cindy replied.

"Let's run it by Eddie and see what he thinks," Beth said.

They found Eddie on the production line and explained the problem to him.

"Why not just inscribe a warning on the outside of the headpiece; something like, 'this side to front', or 'this side up'?" he asked.

"Whatever is easiest? I just want some way to tell them apart."

"Let me talk to someone about it and I will let you know what the difference will be," Eddie told her.

What turned out to be the easiest and the most innocuous was to inscribe an arrow into the headband pointing upward. Eddie showed Beth how it would look and they agreed that it would suffice.

The change in the design was not even noticeable to most of the assembly line workers. The only person affected was the worker who assembled the wiring into a bundle, and the machine did the work. He simply made sure there were no kinks in the wiring, and banded them together.

Satisfied that the design now met the new standards, Eddie went back to the programming function. At least two of them had to work to keep up with the production process.

Clarence called in the afternoon to let them know he had found the people he was looking for. He proposed to fly them to Albuquerque the following morning so they could plan to get them into position at the earliest possible time. He would come with them and assist in whatever way he could.

In the evening, Bill and the others sat down to compile a list of items needed for the vigil. "The one thing we really need is some form of communications," Eddie said. "I think we should get satellite phones for the purpose. The remoteness of the area shouldn't affect them like cell phones where you need to be within sight of a tower to receive the signal."

"That's a good point. I will see about getting them early tomorrow morning. I assume we want them for everyone, including Clarence," Bill said.

"Yes, and I also believe you should err on the side of caution with the supplies, especially food. We have no idea how long this will take, and it will not be easy to resupply way out in the boonies," Beth said.

"We could use a helicopter for that if it becomes necessary," Eddie said. "It might be worthwhile looking into doing the insertion by helicopter too."

"The problem with that is the lack of transportation while we are there. We are going to have to be able to get around a lot quicker than we would on foot, and if we run into some emergency, we need a way to travel," Bill added.

"How about doing the insertion by helicopter and having someone bring a vehicle along the next day?" Eddie asked.

"That would work. I take it you want surveillance in place before we start recalling the computers we have already sold," Bill responded.

"That is correct. The minute word gets out about the recall; we have to be vigilant from that point onward. Another thing we didn't discuss, but there is a distinct possibility that the other contact of Joe's is the woman who talked to Bill on the phone when he inquired about the inheritance. She may or may not have accomplices, but she is obviously involved."

"The people who go on the stake out are going to be very curious about all this. How are we going to handle that?" Bill wanted to know.

"If most of them are ex military, they will be familiar with the need to know principle. We can cloak the whole thing in secrecy and caution them from the start that they will not get any further explanations, but that the job could be dangerous. If we pay them enough, which Clarence has obviously already done, that should be adequate. I suggest you come up with some plausible story, just in case things get nasty," Eddie told him.

"I'm not sure I could come up with a plausible explanation for watching a single spot in the middle of nowhere. The thing we have going for us is that our story is so utterly unbelievable that we could tell the truth and probably not be believed," Bill said.

"Another problem is going to be our rendezvous with Joe. I don't see how we can do that with others watching," Cindy said.

"Maybe we should do that before we bring in the surveillance crew, but I don't see how we are going to be able to do that either," Beth added.

"It may not be a problem," Eddie said. "I don't think we physically leave our bodies during the encounters, so all the people watching will see is the four of us camping in the meadow. Just to be safe, we can have Clarence with them to observe for anything unusual."

"Let's use that as a rough game plan then. It will be Saturday before we can make it up there," Cindy said.

"Clarence will probably bring his plane, so we can have him fly into Fresno and rent a couple of trucks to ferry supplies and people into the mountains. Then on Saturday we can do the same thing. That will give us an opportunity to pick up anything we missed in the original shopping list," Beth said.

"I don't see much more we can accomplish until Clarence gets here with the people. I will go shopping in the morning and try to have most of the essentials before he gets here," Bill said.

Clarence and the five men he had hired arrived before noon the next day. Eddie left Beth and Cindy doing the programming and he met them in the office. Clarence introduced the men to Eddie.

"What have you told them?" Eddie asked.

"Only that it is a long surveillance job at a remote location. Everyone is ex military and all have combat experience. I left the time and payment open for you to discuss with them, but all have done work for me before and are trustworthy," Clarence told him.

Eddie laid out the rough parameters. "The location is a remote part of the Yosemite National Forest. We will

148

need four wheel drive vehicles to get in, and the location where we will set up is in some really rough terrain. I don't know how long the job will last, but my proposal is to pay each of you ten thousand a month, with one month minimum guaranteed. There is nothing illegal about what you will be doing, except camping in an unauthorized area. One of my people will be with you at all times, and if any action is required, other than self defense, he will be the one to act. I assume you are all armed?" he questioned.

All nodded in the affirmative. "Bill, who will be with you, will have a rifle, and I suggest at least a couple of you take rifles as well. If you don't have them, we can purchase whatever you prefer. I would not anticipate a need for them, but you may run into a bear or something like that," Eddie said.

Eddie continued, "Bill is out shopping now. We will communicate by satellite phones, so you won't be totally on your own should you need assistance. I wish I could tell you what this is all about, but you wouldn't believe me anyway, so let's just leave it at surveillance."

One of the men asked, "What about binoculars, or spotting scopes?"

"Bill is out shopping now. I will call and ask him to add a couple of pairs of binoculars to his shopping list."

"I suggest that you fly into Fresno. We can rent trucks there and pick up food and tents. Just plan this like a long camping trip and make up a shopping list for whatever you will need to handle a couple of weeks. We will resupply you as necessary, probably by helicopter. Are there any questions to this point?"

The apparent leader of the group spoke up. "The payment is satisfactory if there is no rough stuff involved. I would like to have a bit more information about what we are watching for and why at such a remote location."

"We simply want to know if anyone shows up at the location. If someone does show up, Bill will decide what to do about it, but it probably will result in the death of whoever shows up. I can't tell you any more, except that what we, and by extension, you, are doing is of vital importance to our species on this planet. There is the potential for wholesale slaughter of a very large number of people if we fail in what we are trying to do. If you know Clarence, then you know that we are not in this for the money. He has more than he will ever need, and my people are not far behind him. While we are on the subject, my group will be coming up there this weekend to check the site out more closely. We should arrive sometime Saturday evening and will coordinate by sat phone.

"I may be preaching to the choir, but take rugged and warm clothing. The elevation is pretty high, and there is the potential for adverse winter weather at any time. Go out and buy what you need. I can give you some cash now or reimburse you later, your choice."

"It probably would be better if you gave us something up front. The bill is going to be pretty large for all of us."

Eddie went to the safe and took out a banded stack of hundred dollar bills. "There's five thousand there. Don't forget the weapons and ammunition while you are out. How did you get here?" he asked Clarence.

"I rented a van at the airport."

Eddie handed the keys to his truck to one of the men. "Take the truck too, so you will have a way to haul what you buy." He then called the gatehouse and told the guard that someone else would be driving his truck.

The leader picked up on the fact that the guard needed to be told to expect someone else to be driving the vehicle. "Are you this careful about everything?" he asked.

"What we are doing here is producing the new computer my company has come out with. It is light years ahead of what's in second place, is proprietary, and is worth billions of dollars. Under those circumstances I don't think there is such a thing as being too careful."

"You are the people who own the company making those machines?"

"We are."

"I had the opportunity to try one of them and it blew my mind. I couldn't conceive of anything operating that rapidly. You think something, and before the thought has time to register in your mind, the picture is on the screen. You don't look old enough to have come up with something like that."

Eddie laughed. "Inventiveness has no age requirements. Our company also developed the drug used to cure AIDS, and our female component came up with the camouflage make up that is so popular now. Clarence can tell you that our potential is unlimited. We try to make our products available to the people who are of limited means, or in other words, affordable. Our philosophy is that our products should benefit the entire population, not just the rich or wealthy."

"That kind of eases my mind about working for you. I have great admiration for the way you do business, and your little speech reinforced my admiration. With your permission we will now go shopping with a clear conscience."

Eddie laughed. "I forgot to tell you that the money is counterfeit."

They all got a good laugh out of that. After the group left to shop, Clarence and Eddie got down to business.

Eddie said, "Bill is buying a bunch of satellite phones today, along with some powerful binoculars, a rifle and whatever else he feels is needed. He is going to stay with the people up there and do whatever he has to in order to keep anyone from contacting Joe."

"Is he going to be there the entire time?"

"Until someone shows up, or until we get all the bad computers back."

"Is there anything I can do to help?"

"If you could go up this weekend and stay with the people watching it might help. We don't know exactly what happens when we are in contact; whether there are any changes to our physical bodies. I don't think there is, but the only way to be sure is to have an observer."

"I hadn't given that any thought either. There is no sensation physically, but the bodies might become rigid or exhibit some other manifestation of the contact," Clarence said.

As they were talking, Bill returned from his shopping trip. "I thought of a lot of stuff we need, but thought it would be better to wait until we get to Fresno to buy it."

"What kind of things are you talking about?"

"We will need a stove for cooking food, and propane to fuel it. I don't think we want to cook over a campfire, and possibly alert anyone to our presence. How did the session go with the troops?"

"About as expected. They are satisfied with the monetary arrangement, but are extremely curious about the reason for the stakeout in the middle of nowhere. They were somewhat mollified after they learned that our company was producing the new computer. They will no doubt try to get more information from you during the vigil, especially if the wait is a long one. Clarence is going up with us to observe while we contact Joe. We were just discussing whether or not our bodies give any indication that something is going on when we are in contact."

"It didn't seem like there was any effect in the past, but we were someplace else and wouldn't know I suppose."

"That's why I would like him present, to get some definitive answer to the question."

"What happens if I have to take care of a visitor?"

"Either bury them someplace in the woods, or call and we will come up and give you a hand. Put a pick and shovel on your list too."

"Can you think of any other way to deal with this?" Bill asked.

"Frankly, no I can't. I doubt that the person working for Joe will know anything about his plan, and could be just like Clarence, a messenger. I don't like the idea of killing someone on the basis of the contact, but the stakes are just too high for us not to do it. Even at this stage of the game, the death toll would be over ten thousand, and the credibility of the computer would be destroyed. That doesn't even take into account what the authorities will do to us for producing the machines."

"What's one life more or less for the cause?"

"If you are having second thoughts I will do it. The computer was my brainchild and I feel more responsibility than the rest of you, so the burden actually falls on me anyway."

"I am not having second thoughts. I can see the necessity, but I have never killed anyone before, and I don't know if I can do it."

"Then you and I should switch roles. I have the same thoughts as you, but I feel more responsible for what will happen if we don't do it. You will be able to handle the production process okay. You can call if you need anything," Eddie said.

"Are you sure you will be okay with this?"

"Yes. I weighed all the pros and cons when we talked about it earlier. It's the only way to assure that we can get all the computers back and one death is better than thousands."

"Do we still plan to go up Saturday and make contact?" Bill asked.

"I think we have to. We need to let him know about Clarence to get his reaction."

Chapter 16

Bill and Eddie relieved Beth and Cindy on the production line for the next couple of hours and Clarence talked with them while waiting for the others to return. They would not be starting the overtime until Wednesday, so when it was time to knock off for the day they gathered in the office again.

Eddie said, "Clarence, you can fly into Fresno, either tonight or early tomorrow morning. All of you go ahead to the location Bill shows you. The girls and I will fly out late Friday and come on up on Saturday. If you think of anything you need between now and then, call us and we will bring it. The satellite phones should work okay from there."

Clarence asked, "Do you want me to send the plane back for you, or will you take care of your own travel?"

"It might be best if you sent yours back. Make arrangements for a four wheel drive crew cab truck for us. Let the pilot know the particulars and we won't have to worry about that later."

"Do you want me to get tents for you guys?" Bill asked.

"Why not, since you will be shopping anyway."

Clarence said, "We might as well get in the air now so we will be ready to go first thing in the morning in Fresno."

"I got half a dozen of the sat phones. That was all he had in stock. I think that will be enough for what we want to do," Bill said.

"Then I think Clarence is right. The sooner you guys get to Fresno, the sooner you can prepare for the remainder of the trip," Beth said.

The entire group went to the airport in the van and Eddie's pickup truck. Clarence's plane was a fourteen passenger model and there was plenty of room for all the equipment they were taking.

Eddie talked to the pilot and set up a time for their departure on Friday night. Bill would call and tell them when the group was in place so they could initiate the recall on the computers that had to be replaced.

The call came late on Tuesday. The group had arrived at the location and everything appeared normal. They found a good place to set up camp within sight of the alien location that could not be seen from the meadow. They camouflaged the campsite and settled in. The weather was already very cold at that altitude, and Bill suggested that they bring additional waterproof ground cloths and blankets. The canvas could be used to construct windbreaks and shield the campfire from view when they had it going. Bill also requested a hand saw and additional axes.

The production line worked the twelve hour days on Wednesday and Thursday. They got an estimate of the time it would take to produce enough to replace the recalls and the following morning Beth initiated the recall procedures. She set it up so the defective units could be shipped directly to them and the replacement would be shipped within a week. If they had to dip into the models for sale, they could do so and just adjust the delivery schedule as necessary. It looked like they had a handle on it, but they were working some very long hours.

When Friday night arrived Beth, Cindy and Eddie met the plane as scheduled and they were in Fresno before ten PM. The truck was waiting for them, and Eddie transferred the additional equipment he had brought along to the pickup. He told the pilot to expect Clarence and the three of them to be back by late Sunday evening, and to be ready to head back to Albuquerque then.

The drive into the mountains took the rest of the night. The GPS receiver Eddie brought along helped them to navigate, in a manner of speaking. It was not like MAPQUEST, but at least kept them going in the proper direction in the darkness. That, along with the forest service map he had, assured he took the proper roads.

They arrived in the early morning hours, just as the sky started to lighten. They could not detect the camp site from the road, and rather than blunder around Eddie called Bill on the sat phone. He came out of the woods and directed them to the camp.

"Where did you guys park the vehicles?" Eddie asked.

"They're back a ways in the trees. You might want to just go ahead and park in the meadow where we will be camping. We can ferry the additional equipment up the hill later," Bill said.

"I don't know if we want to tip our hand just yet. I think it would be best if I parked wherever the other vehicles are and we can take the stuff you requested off the truck. That way when we show up at the campsite we will only have what we need for ourselves. I don't know how perceptive Joe is, but Tom seemed to think he could read between the lines pretty well."

"In that case, follow me. The terrain is pretty rough, but the truck will handle it." Bill then walked back the way he had come and Eddie followed in the truck.

Bill led him around a hillock that was screened from the road by the contour of the land and trees. The only way the vehicles could be seen was from the air. They had a pretty elaborate set up for the camp. The trucks were used as windbreaks and additional canvas was strung from the trucks to trees to provide additional protection from the elements. Folding cots were inside the tents, so the sleeping would at least be fairly comfortable.

Clarence greeted them and the group sat around the campfire and had breakfast with the surveillance group. It was surprisingly good considering the circumstances. Eddie and the girls had slept a bit on the flight, but were pretty worn out from the busy week.

"I don't know about you Cindy, but I need a little rest," Beth said.

"I'm with you. Can we borrow a cot?" she asked.

One of the men pointed to a tent and the two women immediately sacked out.

They slept until noon and Eddie woke them up. "I think we need to change sites pretty soon. The timing would be about right if we left this morning to come up here."

All four of them were apprehensive about the coming encounter with Joe, but knew it had to be done. All their camping gear was still in the truck and they loaded up and headed back to the forest road to get to the meadow. As they started to unload the gear, Joe made the contact. "Greetings. What brings you to this wonderful location this time?" he asked.

They found themselves in the same nebulous location as before, as near as they could tell.

Eddie acted as spokesman for the group. "We are concerned about a couple of things. First is the fact that we confronted Woodman and he wouldn't admit to being in contact with any of your kind. We didn't believe him and decided to turn the tables on him and have him watched. If he is somehow involved in a plan to destroy the earth it doesn't follow that we would do that willingly, because it would assure his destruction as well. So far we have not run across anything that could do what you say. We are going to need more help."

"I didn't say it would be easy. Do you have the new computer on the market?"

"Yes, we have already sold almost twenty thousand of them. People are clamoring for them faster than we can produce them, even with the plant running at full capacity. We have already made almost a billion dollars off them."

"You did a good thing putting Woodman under a watch. I fear he might contact my evil twin and tell him about the computer. If he does, then that might be the way the destruction occurs. Be wary of Woodman if he asks you to meet my compatriot. I imagine he will try to devise some way to use the computer to accomplish the destruction."

"Is that possible? How could he do that?"

"By telling you that I gave you the idea to do it and incorporated something into the design that would accomplish that."

"I'm not sure I understand how he could do that if your kind is not allowed to interfere with us in a direct manner," Eddie said.

"I can give you an example. If he can convince Woodman to bring you to him, he could say that I have in some way influenced you to design the computer in such a way that it could be responsible for the destruction. I don't know for sure, but that is certainly a possibility."

"Do you think we should meet with him if Woodman proposes it?"

"It might be worthwhile, just to see if he does as I think he will."

"So if we meet with him and he suggests that you have some ulterior motive with the computer what do we do?"

"You could come back here and I might be able to suggest some way to deal with the problem."

"We had planned to camp for the weekend, but in light of the circumstances, I think it best that we return and figure out some way to deal with the problem," Eddie said.

"I am confident that you will be able to work it out."

The meeting was much shorter than any of them had figured, and since they had accomplished their goal it did not make sense to spend any more time there. They threw the stuff back into the bed of the truck that they had unloaded and drove away. Out of sight of the meadow, they met Clarence and told him of the meeting.

"Well, I didn't notice anything peculiar, so I guess there is no outward appearance that anything unusual is happening during the encounters," Clarence said.

"He actually suggested that Tom would try to convince you that something was wrong with the design of the computer?"

"Yes, and he even said that Tom would say that he had incorporated something into the design to accomplish the chaotic effects."

"That puts us all into a position of having to choose who to believe."

"I think the deciding factor will be if someone else shows up to make contact with Joe. He as much as admitted that he had someone else working for him when we asked about the inheritance gimmick. If nobody shows up here within the next week or so, then we might have to rethink the problem. I am leaning more toward believing Tom than Joe, so I guess we go ahead with the computer recall until events lead us to do otherwise," Eddie said.

"Do you think we should tell Tom about this?"

"We have to. Whether he is the bad guy or not, we have to try to evaluate him a bit closer. He said he might be able to help me find a way to get rid of Joe, and Joe as much as told me that he would help me find a way to get rid of Tom. This feels like a football game, and we are the old pigskin being kicked from one end of the field to the other," Eddie said.

Bill added, "I think we have to hold off on contacting Tom again until we see what happens here. If someone shows up to try to contact Joe, then it lends more credence to Tom's side of the story. If on the other hand, nobody shows up here, it strengthens Joe's stock a bit more."

"I agree," Beth said. "We are boxed in at this point. We don't know which of them is telling the truth, and our choice of who to believe could decide the fate of the human race, at least to a large extent."

"Let's give it a week with the surveillance here, and if nothing happens we go back to Tom and lay it out and see how he reacts. I think we have to go ahead with the computer recall. There's no way on earth we can stop production of the machines now. People have had a chance to see them in action, and any story we came up with would not be believed. Going to the government is not an option either. I don't think we could convince anyone that we have had contact with aliens, even with the evidence of our exploits over the past couple of years," Eddie said.

"I believe you are right about that," Clarence said. "Whatever happens now is going to be entirely up to us. If we make the wrong choices then the entire population will suffer the consequences."

"How are we going to figure out which of them is telling the truth?" Cindy asked.

"A lot depends on whether anyone shows up here or not," Bill said. "Cindy, you have been skeptical about Joe from the first meeting. Did you get the same vibes when we were in contact with Tom?" Bill asked.

"Now that you bring it up, I didn't. But we were not subjected to the headpiece, except for Eddie. I don't know if that would make a difference or not, but I felt Tom was more sincere than Joe."

Beth weighed in. "And if Tom was the one plotting to do us harm, then it stands to reason that he would have been the one to implant the computer idea. Why not ask him if he will consent to allowing Eddie to disassemble one of their headpieces so he can compare it to what he built?"

"We might be getting ahead of ourselves. The first option is to see if someone shows up here," Bill said.

"Regardless, we are not going to solve the problem right this moment. Give it a week and see what happens here. Bill, it is more imperative than ever that you are able to deal with the situation here," Eddie said.

"I know. I will be able to handle it, but it still goes against my principles."

"Just remember the consequences."

Clarence asked, "Are you talking about taking care of whoever shows up here?"

"Yes. The chances are that it will be a female, possibly with someone else. Whatever we do, we cannot allow them to have contact with Joe."

"I have some experience in these matters from my youth. I don't think I would have any problem helping out with the situation if you would like me to stay with Bill. I would have had a lot of trouble if Tom had asked me to take care of you guys, but knowing the consequences of what we do or neglect to do at this time clears my conscience."

"What do you think Bill, do you want Clarence to hang around?"

"Can you still handle a rifle?" Bill asked.

"Yes. I hunt sometimes, and you remember what goes along with the mental part of our encounter?"

"I had almost forgotten about that. We all have the enhanced physical traits as well. So you can obviously fire a rifle pretty accurately."

"Not as well as forty years ago, but I'm still pretty accurate."

"Then you stay here with Bill. Are there any other rifles in the group?" Eddie asked.

"You've got to be kidding. These guys are all ex military and would as soon go on a trip like this without clothing than without a rifle. Everyone has one, and most of them have handguns as well. We are well prepared, and I accept your offer Clarence. Even the moral support will be welcomed."

"Then that's what we will do. I believe we need to prepare the troops a little better for what may have to happen. I will do that without giving away too much if you guys agree," Eddie said.

"I think you are right," Clarence said. "They deserve some explanation, even if it will not satisfy them entirely."

They went back to where the remaining men were sitting around the campfire. Eddie said, "Gentlemen, we are in a situation where our actions are going to depend on what the other side does. I can't even tell you who the other side is, but I can assure you that you have never been part of any operation that was more important than what may happen here sometime in the next week. Clarence is going to stay and assist Bill if he needs help in taking care of whoever shows up here. We have determined that the only option is to eliminate the visitors right away if and when they show up. Unless the numbers are greater than what we expect, you will not be asked to participate in the action. However, the situation is so critical that if the unexpected happens, you may have to defend yourselves or be asked to help take care of matters."

"You mean you may want us to kill someone?" asked one of the men.

"Yes, if the numbers are greater than Clarence and Bill can handle. We anticipate only one or two people, but there could be more people like yourselves along to provide protection for the main players. If that happens, then the fate of the world literally rests in your hands. I hate to be melodramatic, but the situation is too complicated to try to even explain it."

The leader of the group spoke for them. "We have dealt with situations like this before. We will pull our load if the odds turn out to be more than you expected. With what you guys are worth financially, it doesn't stand to reason that you would be doing something like this for financial gain, or for revenge for that matter. We have been discussing the situation among ourselves and decided that you were the good guys, so we have no problems if the going gets nasty. Just have someone tell us what is required and we will do it."

"Thanks for your understanding. When this is over I hope to be able to tell you more. If it goes wrong the aftermath will be too horrible to even contemplate, so thanks in advance."

Eddie, Beth, and Cindy drove back to Fresno during the afternoon and had the pilot take them back to Albuquerque. They were home by ten at night, and all slept soundly for ten hours.

The hard part for them was going to be the waiting. They would be busy during the week, which would help them keep their minds off the others, but it was still a tough week.

Chapter 17

In the meantime, in Yosemite National Park, Bill and Clarence were consulting with the leader of the paramilitary group to select the best location to take care of any visitors. They had to take them far enough away that Joe would not be able to detect the action, and rather than rely on the information Tom had provided with relation to the distance Joe could detect activity, they decided to err on the side of caution.

Bill said, "I think the best place would be before the last curve in the road. I also believe we will have better odds of success if we can get him or them out of their vehicle before we take the shots. Anyone got any ideas?"

One of the troops said, "Why not block the road with a large boulder, or boulders. It is not all that unusual to have boulders tumble down mountainsides in this kind of country."

"Or even a tree across the road. That happens a lot on the lesser traveled roads in forested areas. If we can find one rotten enough we can probably just push it over."

"Let's go scout out the area and see what will work best," Clarence said.

All of them but one left the camp site and went to look the proposed ambush area over. They first located a good perch for the shooters, then followed the line of sight to the area most desirable for the quarry. The cover was adequate and the line of sight was good. Now all they needed was something to make the vehicle stop and the driver get out.

As it turned out they had the choice of trees or boulders. There were several trees rotten enough to topple onto the road. They chose the largest and two of the men exerted enough force to get it to topple into the road. It effectively blocked passage unless the driver took a very torturous route through the boulders. The only other option was to drive over the tree, which even a high clearance four wheel drive would have difficulty doing.

Satisfied with the roadblock, they used pine branches to obliterate their tracks and those made by recent vehicle use beyond the fallen tree. They didn't want anything to cast suspicion that the tree was anything more than a result of nature.

Clarence and Bill then went back to the area they had chosen for their perch and started carrying the things they would need from the camp. Since they had no idea when people would show up, they decided that the location was as good as any to do the surveillance. Anyone coming to the area would have to travel down the forest road. It ended another two miles down the mountain and it was very unlikely that anyone would choose that route rather than a road, even a primitive one.

When all was in readiness they settled in to wait. It was late Tuesday when they detected the sound of a vehicle in the distance. Bill alerted everyone and they took their predetermined stations. The leader of the hired troops was with Bill and Clarence, while two of the men had been assigned to watch the meadow in case someone should get by. The other two had taken an oblique angle to another location that provided a sight line to the area where the vehicle would have to stop.

Soon the sound grew clearer and the vehicle came into view. It was an all wheel drive Subaru. While it did have good traction, it was not designed for travel in country this rugged. When the vehicle got to the tree across the road it stopped and the driver simply looked at the tree from inside the car for perhaps thirty seconds. The door then opened and a middle aged woman dressed in jeans and a heavy flannel shirt stepped out of the vehicle.

Bill had a rifle with a scope, though at the distance they were shooting he didn't need it. It did provide the confirmation that the woman was armed with what appeared to be a semi-automatic pistol in her belt.

Bill scanned the car with the scope to be sure she was alone, then trained the rifle on her and gently squeezed the trigger. The body jerked and fell almost straight down to the ground. Just to be on the safe side, Bill fired twice more at the prostrate figure.

Clarence and Bill moved toward the car, which was still idling. Bill checked the woman for a pulse, which he didn't find, and Clarence turned the car engine off. There was total silence in the woods.

Bill went through the woman's purse and checked her identification. She had a California address near where his aunt who supposedly left the inheritance lived. He extrapolated that she had been the voice on the other end of the phone when he called to verify the inheritance.

There was no camping gear in the car, so they assumed that she planned on a quick rendezvous and then a drive back out of the park.

They would drive the car to the end of the road and either hide it among trees and boulders, or run it into a ravine. The body would be buried in a different location, hopefully never to be found.

Bill didn't feel the remorse he thought he would. It was simply something that had to be done. Clarence enlisted the help of two of the troops to help find a place to bury the body, while Bill and another of the hired hands cleared the tree from the road and drove the Subaru past the meadow and on to a likely location to ditch the car. One of the trucks followed to provide transportation back to their camp.

The entire process took a little over two hours. Bill had not bothered to call the folks back in Albuquerque yet. They were still watching the meadow, just in case the woman was not the only one with whom Joe had been in contact.

After all the details were taken care of, Bill called Eddie on the sat phone. "Well, a woman showed up and we took care of her before she got close to the meadow. Everything went according to plan, but we are still keeping an eye on the location."

Eddie said, "I have to take this to mean Tom is likely the good guy, but I still don't know how much time we will have to get the bad computers back. There are too many unknown factors at this point. Do you think it is a good idea to leave Clarence there and you to come back here?"

"Let me ask him and I will call you back in a little while."

After he finished the call to Eddie, Bill got Clarence aside and told him Eddie's suggestion about leaving him with the troops to continue the surveillance of the meadow.

"I don't have a problem with that. You guys need to generate some sort of plan about what to do from this point forward, so it makes sense to have you go back."

"Then I will leave this afternoon. I will have Eddie send the plane to Fresno to pick me up."

"Be sure to keep me posted, so I know what you guys decide."

"I will do that, and if anyone else shows up here, you know you will have to take care of them before they can contact Joe. That's probably more important than ever now," Bill said.

"We can handle it here. Just figure out a way to get all the bad computers out of circulation, that's the most important thing right now."

Bill called Eddie again and told him to send the plane back to Fresno, and that Clarence had agreed to stay at the camp with the surveillance team. He then took one of the trucks and drove back to Fresno for the return flight to Albuquerque. He had a lot of time to think on the return trip, but everything came back to figuring the best way to deal with Joe. When he got back to the office he was no closer to a solution than when he started. Maybe the others would have some ideas.

The plane was waiting when he got to Fresno and he was back home before dark.

Chapter 18

The four of them had a strategy session over supper. Beth had stopped by a restaurant and brought home Chinese food.

Bill told them about the woman showing up and how they had taken care of her. "The fact that she was there makes me think Tom is the one who is on our side. I still don't understand all I should about their relationship, especially as it affects us."

"I keep coming back to what Tom said about helping us to deal with Joe. I have no idea what he has in mind, but I also don't have any clue about anything we could do on our own," Eddie said.

"How are the recalls going?" Bill asked.

"We only did the recall a couple of days ago, and already some of the headpieces have come back. Within the next couple of days we are going to have delivery trucks lined up waiting to drop them off. We need to come up with a location to store them that is secure. I'm thinking about hiring someone to do the data entry for the returns at the loading dock. It will be easier that way. We can simply transfer the units to a waiting semi and when it is filled, take them to wherever we decide to store them," Beth said.

"I think we actually want to destroy them rather than put them into storage," Eddie said. "If we do that, there will be no room for error."

"What method are we going to use to destroy them?" Cindy asked.

"Incineration would be best. The wiring will melt and that is the main thing that has to happen. Beth, check into an incinerator we can rent, or even buy if necessary. The best bet might be a funeral home. They all have crematorium these days, and the temperatures have to be pretty high to take care of the bodies."

Beth made a note to do that and the conversation gravitated back to dealing with the alien problem.

"Even with the woman showing up in the meadow, there is still nothing ironclad to tell us which of them is telling the truth. While I agree that her showing up makes me lean toward Tom being the honest one, the stakes are just too high to take anything at face value," Eddie said.

"It would be really ironic if after getting all the computers back and putting out the new one, the original design turns out to be the one that is good," Beth said.

"That thought has crossed my mind more than once," Eddie said. "When Tom showed me the design and what was wrong with it, it seemed too simplistic to do what he said it would do, but I don't have the knowledge to refute his assertion."

"Maybe when we go back there you can ask him to show you the design of the one they use and see if you can pick out any other differences that might possibly come into play."

"What you have to remember is that Tom is the one who told us that Joe was up to no good. Otherwise we would not have any idea that anything was wrong with the design," Bill said.

"But at the same time, that could have been a ploy to get us to change the design for his purposes," Beth said.

"Let's try to get back to Tom this weekend and see if he is willing to shed more light on the situation. It is maddening not knowing if we are taking the right action or simply adding to the problem with the design changes," Eddie said.

"Do you think you would know if he shows you the real thing? What I am getting at, is that he might be able to have you see what he wants you to see, whether it relates to the actual design or not," Bill said.

"I would know if the design is the same as mine. Whether or not that will be helpful in determining which of them is truthful is a horse of a different color. There is not a lot we know about them or their capability to influence our knowledge. They both say that they simply made us capable of using all our brain power instead of the small portion the normal person uses. That could be true, but it could also be that they used the devices to implant the extra intelligence into our minds. I certainly don't feel a lot smarter in trying to figure this problem out," Eddie said.

"We definitely know that a change occurred in all of us, including Clarence, who preceded us by forty years. The fact that Clarence was not asked to do anything but watch for someone like us over such a long period is another point in Tom's favor. He could just as easily have asked Clarence to take care of us permanently," Cindy observed.

"I don't know if they are supposed to have personalities, but Tom seems to have more social graces than Joe. That certainly is another point in Tom's favor," Beth added.

"I am to the point that I think it would be best to go to Tom and express our frustrations with the situation. His reaction might be enough to help us decide which of them is on the level. I am willing to try anything within our ability to get this right. The stakes are just too enormous for us not to explore every avenue to make sure we are right," Eddie said.

"With the headpiece they use it seems they can read anything in our minds. Why not have Tom scan all of us and see if he can come up with a method of discovering the truth within our own brand of reason. He got the information from you very quickly when you used the headpiece, so he will be able to do the same with all of us and show us the logic pattern," Beth said.

"The other side of that argument is that he could implant a subconscious thought in each of us to place the suspicion on Joe, if in fact he is the bad guy," Bill said.

"I keep coming back to his offer to help neutralize Joe. If he gives us a method to do so, he is in effect telling us how to get rid of him too, and I don't think he would do that," Eddie said.

"Let's call Clarence and set it up to pick him up after we close down on Friday. Remember, the group up there only has the one vehicle now, and we will need to ferry a vehicle up for them to accommodate everyone. We can do like we did last weekend, except we take two trucks and leave one with them. I assume we still want to watch the location, and I don't think the troops up there will have any qualms about taking care of anyone who shows up after what they witnessed with the woman," Bill said.

Eddie picked up the sat phone and called Clarence. The conversation was not a long one. Eddie reiterated what they had decided and told him they would be there early Saturday morning. "The one thing you need to do is assure that the troops are willing to take care of anyone else who shows up. We don't know if Joe had additional people or not, and he could still do a ton of damage if he finds out what we are doing."

"We have discussed the situation and I sweetened the pot a bit in case they have to do anything other than just watch. Earl, the leader, is okay with it, and I get the feeling it will not be a new experience for them," Clarence replied.

"We will fly directly to Philly from Fresno if that is okay with you. I will tell the pilot to get his rest for the follow-on flight while we are on the road."

"That will work. Is there anything else I need to know?" Clarence asked.

"I can't think of anything. The headpieces we recalled have begun to arrive, and we decided to incinerate them to make sure they are totally destroyed."

"Then I will see you when you get here," Clarence said.

The remainder of the week was hectic, with the overtime to produce the replacements for the recalled equipment. They didn't have much time to discuss the problem further.

Beth had found someone to log the data into the computer for the units returned. She had been right about the delivery trucks being lined up to make deliveries. More than four thousand of the units came back before they were to leave for Fresno on Friday night.

Beth had lined up the use of the incinerator at one of the funeral homes and Bill and Eddie had burned the first load of the headpieces. Eddie commented that they may have to replace the incinerator for the funeral home because a lot of debris was left over after things melted down.

The system Beth had set up for tracking the returned units included listing the names in a program separate from the sales files. She could then cross reference the returns and delete the names on the sales list as they were confirmed by the first list.

When they finally got ready to leave on Friday evening, all of them were exhausted by the week's activity. They all slept for the short flight to Fresno. Eddie instructed the pilot about the coming trip and asked him to file the flight plan and get some rest before morning.

The trucks they had used previously were parked in the rented hangar and they loaded up and headed for Yosemite with a couple in each vehicle. It was difficult for the drivers to stay awake, and they switched off with the driving chore. Again it was the early morning hours before they arrived.

Nothing else had happened during the week; at least nothing concerning the meadow, but a bear had happened upon their camp and for a time it looked like they might have to shoot the bear. He finally ambled away of his own volition, but the group kept a watch around the clock just the same.

Watching for the bear to return was more taxing than watching the road and meadow.

They all had something to eat after their arrival and were ready to get back on the road within an hour.

Clarence gave final instructions to the troops who would be continuing the watch and as the sun began to take the nighttime chill out of the air, they departed. It was a bit crowded with five of them in one truck, but they were so tired that all slept, with the exception of Clarence, who was driving.

By noon they were back in Fresno and loaded on the plane to head for Philadelphia.

Chapter 19

When they arrived they again rented a van. All had caught up on their sleep during the flight and were somewhat refreshed.

It was late evening when they got to the location to meet Tom. As usual, there was no wasted time when they entered the grove of trees.

They found themselves in the presence of Tom once more. By way of greeting he said, "How are events progressing?"

"To be absolutely truthful," Eddie said, "we are more confused than ever."

"What is the source of your confusion?" Tom asked.

"We don't know whether to believe you or Joe. We went back and contacted him again and told him we had confronted Clarence. He then suggested that you might take the actions that you did already, though he didn't know that we had contacted you. He said you would react pretty much as you did, and that it would be your way to get us to modify the headpiece to cause the problem. It sort of seemed logical to us, at least it caused us to have a more critical look at all that has transpired," Eddie told him.

"That is about the way I would expect him to react. He figures I will attempt to set the matter right and wants to cast doubt on my intentions. I don't know how else to convince you that he is the cause of the problem."

"One thing that happened that caused us to look at you more as the good guy was that a woman showed up at the location, probably to report to Joe. Regrettably we had to neutralize her before she could make contact. If she was, in fact, preparing to make a report on our activities, it could have spurred Joe into taking action before we could destroy all the recalled units."

"So how do you propose we solve your dilemma?" asked Tom.

"Would you be willing to let me have a look at the design of one of your headpieces to compare with what I built?" Eddie asked.

"I can implant the design into your mind, but you will have no way to know if it is the actual design, or something I am using as a subterfuge."

"We also thought that maybe if you looked at all of us, you could point us in a logical direction to solve the entire problem."

"I will do that of course, but I can't make any promises about setting you on the proper path."

"Then let's do that first."

Each of them put on a headpiece for the few seconds it took for Tom to do whatever he needed to do.

"First I am going to show you the design of the headpiece," Tom said.

Eddie immediately had a vision of the inside of the headpiece, along with what he knew as a blueprint for the design. He mentally compared the design to what he had come up with and could see the similarities, but he could also see a lot of differences. He had no idea what the additional components did, but assumed they were for functions with which he was not familiar.

Tom told him, "This is the complete design. There are functions that you are not familiar with in the components different from your design."

"Well, the one thing it tells me is that Joe had to have implanted the design in my mind. There's no way I could have conceived something this advanced."

"Don't sell yourself short. He may have implanted some information, but you did the design on your own."

"Okay, now to the second part of the problem. See if we have the necessary information to solve the problem," Eddie told him.

Tom took a bit more time to answer. He was apparently assimilating the data from them collectively.

Finally he said, "You all have the necessary information to solve the problem, but I don't think you will be able to do it on your own. Too much of the information conflicts with your precognitive concepts. What I mean to say is that the information could lead you in different directions and either solution would seem logical to you. Your dilemma now about which of us is acting in your best interest is a good example. A case could be made for either and the logic would fit what you expected to see."

"We all agreed that the one thing that would convince us in your favor would be if you could tell us how to get rid of Joe, or at least how to approach the problem. The reasoning is that if you did that and were trying to deceive us, then we could use the same method to get rid of you," Eddie said.

"I applaud your logic, but if I wanted to deceive you, I would tell you how to get rid of Joe, and then move my portal to a different location that you would not be able to find. Having said that, I will help you get rid of him. It will not be in the sense that he will be destroyed, but his portal will be destroyed and he will have to return to his ship in order to reestablish contact with your planet."

"Okay, how do we go about it?" Eddie asked.

"It will take a series of four electromagnetic oscillators, positioned at precise locations around the portal. The oscillators will have to be at the same output, and be pointed so the beams meet in the circle where you made contact. The units will have to be at least fifty yards from the circle to remain undetected. You will have to build the oscillators yourselves since they do not exist on your planet."

"How big do they need to be?" Eddie asked.

"I will implant the design in all of your minds, including Clarence. You will need to decide how to do the logistics, and I fear you will also need to coordinate with some of your government people to get this done. I cannot help you with those aspects. Here is the design of the oscillators," Tom said as the design flashed into each of their minds.

"Whoa," said Bill. "These things are huge."

"That's what I meant when I said I couldn't help you with that part of it. There is no way you could get four of these to the location without someone finding out about it."

"How long do you think it will take us to build these," Clarence asked, "and will they be of use to any of our industry or military operations?"

"They could be used to generate electricity, or even as a sort of electromagnetic pulse weapon. You might be able to induce the military to help build them if you can show them a use for the units after you finish using them."

"You are sure this will work?" Eddie asked.

"Yes, but you need to get all the bad headpieces off the market before you start on this," Tom said.

"You said a minimum of fifty yards away from the portal. Is there a maximum distance?" Eddie asked.

"You don't want to be miles away, but you could be as far away as a quarter of a mile, as long as all the units are at the same distance."

"Well, for better or worse, you are now the good guy. We may be back for additional instructions when we get into the design."

"I am always at your disposal," he said.

Each of them removed the headpieces and left the grove of trees. As they walked back to the van Clarence said, "I am not mechanically inclined, but I think I understood all that. The hard part is going to be dealing with the government. That might be where I can be of the most use. I know a few people in influential positions. Some are elected officials and others are political appointees. Where should we start?"

"I say go right to the top. The President is going to have to be briefed and I would rather we do that than depend on a go between. Things sometimes get lost in translation, and I believe we are going to have to reveal a bit more about our situation to convince him to devote the manpower to this scheme. We will put up the money between us, so that shouldn't be an issue."

"I will start working on that right away. I know I am not going to be able to present this on my own, so what I would like to do is have a private meeting with all of us present if I can get him to agree to it. I will make the cocktail circuit around Washington and see if I can make informal contact. He knows my name, and that I am quite wealthy. That should be enough to at least get a meeting with him. I can then give a broad brush outline of a potential new weapon and try to get him to agree to meet with all of us."

"I like that idea," Beth said. "I think if we gang up on him, we will be able to convince him that he has nothing to lose by cooperating with us. Also the potential of what we are proposing will assure him of reelection, and that's a carrot he can't refuse."

"We are going to have to work with the military to get the recalls of the computers we sold them, so we might be able to make some discrete inquiries to senior officers about the potential for a new weapon. The computer is enough to convince them that we are serious, and we will need to identify a location to build the things," Eddie said.

"Let's go get something to eat. All that outdoor cooking makes me yearn for a decent meal," Clarence said.

They found an Italian restaurant along the way and went in to eat. As they munched on salads, they continued to discuss the different aspects of what they would have to build.

"Each of those units is going to weigh about twenty tons, and the remoteness of the location is going to make it very difficult to get them in place," Bill said.

"We might have to build them in sections and transport the sections separately. I don't know what the lift capacity of the large helicopters is, but it would be nice if we can keep the weight under their capacity so that we can transport them that way," Eddie said.

"We are also going to have to keep our surveillance in place until we complete this project. Once we get the okay to build it, maybe we can get that particular forest road closed off," Clarence said.

"Maybe we won't have to wait until we get all the recalls back before we start. Some of the components, especially the housings are going to have to be fabricated by someone else. It would be cost prohibitive to do that on our own. We can do the blueprints with the new computer and take them to some metal fabricator and have them ready by the time we are ready to start with the rest of the design," Eddie said.

"Whatever we do, it probably should be close to Yosemite, either in Nevada or California," Beth said.

"That is something else we can get a head start on. If we find a location, we can go ahead and acquire the space to build them. How about you and Cindy start working on that between our shifts at the plant?" Bill asked.

"Sure. We can get on that right away. The more pieces we have in place when we are ready to start, the better off we will be," Beth said.

"The way I understand what we were shown, is that the negative force will be projected to meet in the location of the portal. This probably forces anything under its influence upward, since that is the path of least resistance. I am assuming that causes the portal to collapse," Eddie said.

"That stands to reason. It doesn't matter how it reacts, as long as it gets rid of Joe," Cindy said.

"Do we all agree that this is the action we want to take?" Eddie asked, just to be sure they were acting in concert.

Each of them indicated that was the case, so the die was cast now. Whether they were right or wrong, only time would tell. That was assuming they could get the oscillators built and in position to use them without anything going drastically wrong.

"Beth, how long before we get all the original headpieces back?" Eddie asked.

"We should have most of them inside of two weeks. There are going to be some that will be harder to locate, especially those in use by the military. What we don't have by that time will only be a small percentage; numbers wise likely less than two hundred."

"That's still a lot of lives to lose, so we will have to make a concerted effort to get the military to cooperate with us. If we have to, we will hand deliver them," Bill stated.

By the time the meal was finished, they had a pretty good idea of what they needed to do, and the timeline for accomplishing the tasks.

"I don't see much need to go back out west," Clarence said. "I am going to just turn the plane over to you guys, and if I need to travel I will charter a plane or fly commercial. I only see the need to travel between Washington and New York anyway."

They continued on to Philadelphia and turned the rental van in. They then flew to Washington to drop Clarence off and continued back to Albuquerque.

Chapter 20

Now that they had a direction for their efforts, all felt a bit better. Eddie used the computer to pull the details of the design for the housing of the electromagnetic units they would have to build from his mind. He then printed them. He looked them over carefully to get a better idea of the scope of the effort. He also wanted to see if there was a way they could build them in sections. It appeared that the design was a single unit, but that it could be built in two sections and then welded together. He didn't see a need to include the access panels that would be required if the system was used only for their purpose. If they had other applications at a later date, then the doors could be added.

As he looked at the housing plans and thought about the design in his head, he visualized the way the things would work. The front was open in the center of the housing and an apparent aiming device protruded, similar to the barrel of a large artillery piece. Since the unit was too large to manipulate once it was in place, they would have to make sure they were lined up properly during the building process. The aiming device seemed to be of a different material than the rest of the components, probably a non conducting material like fiber glass.

The working parts of the design reminded Eddie of an alternator on a car. There was a lot of wound copper surrounding a central core composed of iron, nickel, and magnesium. At the center was a vacuum tube, which purpose Eddie did not see at that time.

Eddie saw additional problems getting the materials for the core of the units. They would all have to work together to line up the materials they needed.

It was Monday morning and Bill and Cindy were on the assembly line doing the programming. Eddie and Beth would relieve them in three hours.

The unit sales were continuing and they were producing in excess of two thousand units a week, plus the ones they were making in the evenings to replace the recalls. Their total sales were over two billion dollars.

Not a day went by that the major news networks and newspapers did not have feature articles about the computer, and/or the people that designed it. They were besieged with requests for interviews by the major news outlets. They had not granted an interview since the announcement that the computer was to be marketed, and the follow-up for the recall of the defective headpieces.

Since none of them were well known, not a lot of people recognized them when they were away from the plant. The security around the plant was like a military compound. With the fences, razor wire, guards and electronics it was almost impossible for anyone to breech except from the air, and they weren't worried about that aspect.

Clarence called on Monday evening and told them he had talked to some people on Sunday night, and that he might possibly get an audience with the President sometime during the week, though he had not heard anything officially. He wanted to know how everything else was going, so Eddie brought him up to date on the blueprints for the housing and the scope and diversity of the elements they would have to procure for the working parts of the accelerator.

"Do you have any contacts in industry that might be helpful in that regard?"

"I can probably find someone to build the housing, since it requires nothing special. The way I understand it, that is simply a metal housing for the rest of the unit."

"That's the way I see it too. I have printed the specifications out, so if you know anyone who can do the job, put me in touch and we can get a start on that part of the design. The working part is going to be rather more difficult to build. It will require access to metallurgy and probably an industrial forge to meld the metals. That process may require another trip to see Tom, but I won't know until I get the specifications on paper and study them."

Since the design was in Clarence's head too, he asked, "What's the purpose of the vacuum tube, or whatever it is in the center of the unit?"

"I am not sure at this juncture what is in the tube, but it has to be what conducts the magnetism and allows it to be channeled. I assume that it can be either positively or negatively charged."

"With the amount of copper you are going to need, it might not be a bad idea to start looking for sources," Clarence said.

"You're right about that. If the winding is a different size from what you normally find in an armature, we may have to have it specially run. I will let you know in a few days."

After they hung up Eddie started to make notations on a pad for the different things they would need, and the sizes of the elements. The copper wiring to be used for the windings was not a size he was familiar with, but he had not worked with it enough to know if industry made that size or not. The individual mineral elements and quantities he tried to nail down to a good estimate so they could at least locate a vendor that carried the materials.

When time came for Beth and him to relieve Bill and Cindy, he told Bill what he was doing and asked him to continue the list and try to determine if there was anything unique in the system design that they might have problems with.

Beth and Cindy helped with the list, and also spent some time on the phone trying to locate a likely place to actually build the units when they got to that point.

Cindy came up with a likely location in Reno, Nevada, which was not terribly far away. She also found another place in Carson City, Nevada, which was south of

Reno, and if they made the delivery by helicopter was a straight shot across the mountains. She told both parties that they would like to examine the properties, and that they would be in the area over the weekend. She would call back no later than Friday to set up an appointment with the agent.

On Thursday Clarence called and said he had the name of someone who might be able to fabricate the housings for them. He gave Eddie the name and phone number. Eddie immediately called and talked about what they needed.

"From what you tell me, I don't think we will have a problem with it. I need to see the engineering specifications to be sure though."

"I will fax them to you after we finish our conversation. The one thing I need you to do is build them in two sections so they are not too heavy to be moved by helicopter. As long as all the attachment points are built in, we can do the welding to put the two parts together. Also, do not attach the access doors. Simply leave them open and we will take care of that later as well. You will need to add attachment points on the outside so the cables can be attached and the unit will be balanced for flight. I will mark up the drawings with the changes and fax them to you."

"Do that and I will call with any questions and give you a price. When do you want them finished?"

"The sooner the better, but don't go to overtime to get the job done. We have a great deal of work to do on the stuff that goes inside, so time is not a critical factor."

Eddie started to mark the drawings up, then decided that he would use the computer and visualize the design with the changes and print a clean copy. He did that and laid them side by side to make sure he had not missed anything, or screwed something up in his mental translation of the specs.

It looked good, but he wanted Bill to check it before he sent it and relieved him on the production line so he could do it right away.

"After you look at the new drawing, if it is okay then fax it to the number on the pad. If there's anything I messed up, make the changes before you send it."

"You've got it. Man this is tedious work," he said as he departed.

Cindy called the real estate agent in Nevada on Friday and made appointments to see both the properties she had talked to them about. They flew to Reno late that night and stayed in a hotel.

The agent met them and showed them the property in Reno. It looked adequate to the job, but was not very accessible for a helicopter. They next went to Carson City and immediately recognized that it was the superior location. The building was large, with hangar type doors, and the fenced property had ample room for landing a helicopter without any encumbrances, such as electric lines or large trees nearby. They immediately told her that they would purchase the property rather than rent it if the owner was agreeable. The property encompassed maybe ten acres and with the fence and building was probably worth close to a million dollars.

"The owners have not given any indication that they want to get rid of the property, so I don't think they will look on an offer favorably," she said.

"Would an offer of two million dollars influence their outlook?" Eddie asked.

The agent did a double take. "You must want this property awfully bad. It is not worth more than a million."

"Then you think they will accept an offer of two million?"

"I feel sure they will."

"Why not give them a call and find out. The property is ideal for our purposes and we really don't want to spend a lot of time looking."

The agent made the call and told the owners what had been offered. They accepted the offer and just like that, the deal was done.

"We will leave the corporate information with you. When the paperwork is ready, let us know and I will have the money transferred to whatever account you desire," Cindy told her.

"You're going to pay cash?" the agent asked.

"Is that so unusual?"

"For that much money it is."

"If you would prefer us to finance it we can certainly do so."

"No, cash is not a problem."

After they signed the paperwork and the agent had everything she needed for the deal, they went back to the Reno airport and took the plane back to Albuquerque.

They now had a place to start stockpiling the materials they would need. Also, the special machinery could be purchased and held there. Although they knew of no obvious threat, they decided to have a good security system installed and to have the place physically guarded as well.

"Who wants to take on the job of security for the place?" Eddie asked.

No one volunteered right away so he said, "I have an idea. Let's pick the best we have on the security force at the plant and put him in charge of designing and implementing a security plan for the place."

The other three immediately agreed. None of them wanted the extra work, and it was more a precaution than anything else.

With that chore out of the way, they spent the rest of the weekend going over the plans again. The overall plan was to have the casings built and shipped to the Carson

City location. They would take one into the building to use as the prototype, making sure all the components mated as they were supposed to. The remainder could be left in the large lot. They were so large and bulky that no one would attempt to steal them, and with security in place that was not a worry anyway.

"We need to determine the contents inside the vacuum tube and the composition of the materials. It may be that we have to have special machinery for that, or even design something ourselves," Eddie said.

"I wonder how Clarence is doing on the political front," Beth said. "I am beginning to believe that we can do this all on our own if necessary. We might have to buy a helicopter, but we will have to involve a lot of other people if we do that, and I am not sure we can keep the project from Joe with that many people around without some discipline on their movements. The military can be told to avoid the area and they are likely to pay attention. They are not normally given a reason, just an order."

"At least we have gotten a lot of the recall units back. I think after next week we will have to start tracking the remainder down. Some people are going to be stubborn about returning them because they can see no fault in the way they operate," Cindy said.

Beth added, "At least we can do away with the extra shift work after next week. I think we have enough on hand to take care of the rest of them already."

Eddie called Clarence to bring him up to date and find out if he had made any more progress getting an audience with the President.

When he had him on the phone he explained what they had done and where the building was located. "Have you made any progress getting an audience with the President?" he asked.

"No direct contact yet, but I have asked one of the more influential congressmen to suggest that he might want to talk to me. I have been staying in the Washington area, except for a short trip to New York on Thursday. I am hoping he might call this weekend."

"I think we might want to start acquiring the materials we are going to need now that we have a place to store them. I am going to use one of our security people to set up the security there. He will be on it Monday."

"If I hear anything about the meeting I will call you. I think the President will be curious enough to give me a call if the Congressman asks him."

"If you don't hear anything this weekend, I will try to get to a military officer with the idea of developing a new and unique weapon for their inventory. That might be enough to do the trick," Eddie said.

"I really think I will hear something over the weekend. We can still go ahead on our own to get started. It is going to take us quite a while to build these things," Clarence said.

"Yes it will. And it might entail a couple of more visits to consult with Tom. I don't have any idea about how to even start building the vacuum tube. It wouldn't surprise me if we have to go to a physics laboratory to do that."

"If we do, it would be a lot easier to accomplish with the government's backing," Clarence observed.

"You're right about that. We will talk to you again tomorrow night unless you hear something sooner."

After he hung up, Eddie said to the others, "Clarence hopes to hear something from the President tomorrow. If he doesn't I think it's time for us to take a different tact."

"The General who ordered the computers for the government might be a good place to start. He knows our product, and also knows that it is cutting edge technology. He is smart enough that if we give him a rough overview of the thing we are going to build he will see the military application for it."

"Let's call it a night. I might even sleep in tomorrow," Beth said.

As they started to their bedrooms Eddie's phone rang. He checked the caller ID and said, "It's Clarence."

"Well, either you forgot something or you got the call from the President," Eddie said.

"I got the call. It was not a very lengthy one, but was very fruitful. After I explained to him that it involved you guys and reminded him that you had developed the computer, he agreed to meet. He wants us to go to Camp David tomorrow for lunch. We need to be at Andrews AFB by ten thirty tomorrow. A military helicopter will ferry us to Camp David."

"Then I guess we had better fire up your plane and head east. Where are you staying?"

"At the Hilton. Give me a call when you get in and I will have someone meet you and bring you here."

The group packed again after calling the pilots to let them know to file a flight plan. "Suits this time I think," Eddie said.

They wouldn't need much in the way of clothing, since it was going to be lunch and then back to Albuquerque, so the packing went quickly.

They were in Washington in the wee hours of the morning.

Chapter 21

Clarence met them at the hotel and they had breakfast. It was still only seven A.M. and they had a lot of time to kill, so they talked about their approach to the problem.

"How do you see this meeting playing out?" Clarence asked.

"I think we tell him that we have a project that we are working on that might be of interest to the military. If he shows an interest then we lay a little of the reasoning out to him, without telling him about the aliens of course. We will pitch for the military to support and provide security while we are developing it and we will do a full up test to make sure it operates the way we want it to. The test is going to have to be at an isolated location, and we have already decided where that will be, but will not reveal it until we are finished with the development process. Do you think he will buy off on that?" Eddie asked.

"I don't know. I have never met him, other than to shake hands at some social function," Clarence told them.

"Then I guess we just take it as it comes. How do you think he would react if he knew the truth about all of us?"

"I think he would find it hard to believe, but on the other hand, the results speak for themselves. In fact, that might be the best way to convince him that we need to do this, and he needs to be involved. Let's see how he reacts to the sanitized version before we take that leap," Clarence said.

"What's the down side to telling him? If he doesn't believe us, then he is not going to spread the tale around because it will make him look foolish. If he does believe us, then we have an ally who can provide the things we need to get the job done, and the end result will be a weapon the military can certainly find a use for," Eddie told them.

"One way to convince him might be to demonstrate how we interact with one of the computers. The speed with which our thoughts progress will blow his mind, and he can use the computer himself and do a comparison," Bill said.

"Okay, let's take one along in case we need it," Beth said.

They talked for over an hour and pretty much decided that if they could not convince him any other way they would take him to visit Tom. That was a choice of last resort, but could conceivably be done.

Clarence had gotten rooms for them and they went to change clothes. They went by the airport to retrieve a computer from the plane before going to Andrews. They were still early and had to wait for the helicopter to arrive. It was flown by the Marine detachment and was the back up for the President's own bird.

There were Secret Service people at the embarkation area and they scrutinized the computer thoroughly before allowing it to be loaded onto the helicopter.

When they got to Camp David, the President came out on the porch to greet them. One of the Secret Service people from the President's detail carried the computer inside.

President Kelly greeted them. "I have met you briefly Mr. Woodman, but I have not had the pleasure of you four," he said.

Eddie was closest so he made the introductions.

"Well, come on inside and we will have some refreshment while lunch is being prepared."

He led them inside the cabin and to a sitting area large enough to seat them all. A steward brought coffee and condiments. The President said, "If anyone wants something stronger, just say the word."

Nobody took him up on the offer and they all drank coffee.

Eddie noticed the Secret Service man stayed in the room and glanced again at him, hoping the President would take the hint and not make him ask for a private audience. Fortunately he did. "Lamar, you can wait outside if you don't mind. These gentlemen and ladies are too rich to be assassins," he said with a chuckle. "And I have a feeling they want to talk about something very private."

The President had a very disarming way about him, and though he could be tough when the occasion called for it, he was a very charming person.

When everyone else had left they looked to Eddie. They had not discussed who would do the talking, but the President picked up on the glances. "I assume you have been elected spokesperson," he said.

Eddie took a deep breath and started. "You obviously know about the computer we developed," he said, waiting for a nod or reply.

The President nodded. "I have seen them in operation and it is so far beyond anything anyone else can even imagine that there is simply no contest."

"Did you know that Bill here developed a cure for the AIDS virus, and put it on the market at a cost almost anyone could afford?"

"Yes, I know about that too, and let me say that is one of the most generous and humane acts in the history of mankind. But I still don't see where this is going."

"We have joined forces with Mr. Woodman as kindred souls sometimes do. Our combined wealth is almost as great as most countries national budgets, so what we are going to talk to you about is not related to money, though it may seem so. What we want to talk to you about is another advance in technology that is so far removed from today's knowledge that it couldn't even be conceived by the most forward thinking military genius."

He continued, "We are going to develop an electromagnetic oscillator that can be used to generate electricity, or it can be used as an EMP weapon for directed controlled use. Would you be willing to commit some military to the project on our word, or would you take too much political flack for that to work."

"I'm not sure I understand what you just said. I got the part about a weapon of some sort that is highly advanced, but I don't understand the rest of the pitch."

"With or without your help, we have to develop the EMP accelerator, for lack of a better term. The reason is to save untold lives, and take care of a threat to the world at large. I'm sorry if this sounds disjointed but it is hard to give you specifics without getting into something you will not believe. Let me instead give you a demonstration."

Eddie went to the computer and plugged it in. "Have you had the opportunity to use one of these yourself?" he asked.

"No, only a demonstration."

"The way this works is that the computer reads your brain waves and responds in a way that the brain can read the response. It operates at the speed of thought. The instant you think of something, it appears on the computer screen. The images remain until you shift your thoughts to some other subject, and immediately the computer picks that up. I would like you to try this if you don't mind."

He handed the headpiece to the President and showed him how to put it on. Once he did, the computer cycled through the frequencies until it matched and the picture appeared on the screen.

The President kept looking at the screen and as his thoughts changed, so did the screen presentation. "I saw this demonstrated, but I had no idea of the complexity and speed of the thing. I can see why everyone is going bonkers over it."

Eddie motioned for him to take the headset off and then put it on his own head. "Now watch," he said as the images started to cycle through the computer so fast that they barely registered as separate thoughts.

"What's going on now," the President asked.

"You are witnessing the speed at which my thought process works. Each of these four can do the same thing. We actually had to revise the design to slow it down enough so we could use them. Do you want to see the others use it?"

"I will take your word that they can do the same thing, but I don't see the point of it."

"Come now Mr. President, you are much more intelligent than that. Are you telling me that you don't understand the implication of the fact that my thought process is probably a magnitude of a hundred higher than yours?"

The President was silent for a few seconds. Finally he said, "I can see no earthly reason for the difference to be that great. Even if you were a genius, which I am sure you are, the difference should not be that great."

"You said the magic word, earthly. We had hoped to be able to convince you to cooperate with us on the face value of the potential weapons system, but I don't believe that is going to work. What you are going to get now is the unvarnished truth, and let me warn you ahead of time, you will find it very difficult to believe."

"You have heard of the Roswell story back in the 1940's, and for all I know have reviewed the files or artifacts if they exist. The subject of aliens from different planets has been a topic of great interest for the last hundred years at least, but nobody had been able to show proof that they actually exist. You just saw the proof, though you could probably try to explain it some other way. So do you believe in aliens or extraterrestrials?"

"You seem to be saying that you have had encountered them and super intelligence is the result."

"That is indeed what I am saying, but the situation has so many twists to it that we five can hardly believe it

195

ourselves. Clarence here had his first encounter over forty years ago. He was endowed with the ability to use the full capacity of his brain, just as we all eventually were. The story unfolds in stages, so I have to set the stage properly. When Clarence was contacted way back then, he was simply asked if he wanted riches beyond his wildest dreams. The answer of course was yes. The alien said that the only thing he had to do was contact him once each year and let him know if anyone came along that exhibited the same intelligence traits as he now had. Clarence readily agreed and stuck to the bargain for the entire time. You do know that Clarence is in the top ten of the world's richest people? We four may pass him within the next six months, but I digress."

"You see, Clarence had no idea why the alien wanted him to do what he had requested, but had a very hard and rough life to that point, so figured he had nothing to lose. As his wealth grew and his old habits fell away, he started to worry about what would be asked of him if and when he found what the alien had asked him to look for. Then just over three years ago, we four had our encounter, not with Clarence's alien, but a different one. We were told that we had been chosen to save the world from sure and total destruction, but that we would have to figure out how and when in order to stop it. The alien said he could not interfere with the development of different planets, which we believed at first. He placed a device similar to the one used with the computer over our heads and said that now we had full use of our total persons. It was so quick that we about decided nothing had happened. We felt nothing, no ill effects, no elation, no sudden impulses; nothing out of the ordinary at all."

"We had been sitting around a campfire at the time, and when we found ourselves back there, someone said, 'was I dreaming, or did something happen to all of us'? If it was a dream, we all had the same one, so we tested what he had told us. Each of us read a paragraph from a book in turn and then closed the book and repeated what we had just read verbatim. It convinced us that we had indeed had the same experience. Now we had to figure out what this meant. We had been given no clues, except that the world was closing in on total destruction and we now had the ability to avert that catastrophe."

"If I'm boring you, just say so?"

The President said, "You still have my attention."

"Well, we started a small company with some seed money provided by an aunt of Bill's that he hardly knew that had supposedly been left in her will. I checked into what was going on in the technology world and found a company that was working on a processor that would speed the computer process up by a magnitude of ten. They had not solved the chip problem and the solution was in my mind as I read the description. I arranged to go to work for them on a contract basis to develop the chip for them for a set fee of half a million dollars."

"We loaded up on company stock, and I developed the chip for them and integrated it into their system. We made enough money off that for Bill to develop the AIDS cure, and for Beth and Cindy to develop a unique make-up that made over a million dollars in its first six months. We were still obsessed with the end of the world thing and went back to contact the alien to see if we could get more direction. I can see that this is getting too long winded, so let me fast forward to the present time."

"The idea for the thought computer had been bugging me ever since the first encounter we had. I worked on the concept for almost two years before getting it all figured out, and still had a really tough time working out the frequencies at which the machine interfaces with the brain. But eventually it all came together, and the result is what you just experienced."

"The alien had told us that there was one of his kind that was going to be responsible for the destruction and that he probably already had someone in place. He told us what to look for and we immediately picked up on Clarence as the potential adversary. He also picked up on us and was having us watched. We caught onto the watchers and installed a video spy camera in the headquarters they were using, just to keep track of them."

"Eventually we got to the point that we decided to have a face to face with Clarence. We couldn't see him acting in a capacity that would surely end in his own death as well as the rest of the population. When we talked he felt the same about us, that is that we were not bad people, and the things we were doing with the AIDS drug for example, were more helpful than harmful to society as a whole. We joined forces and decided to contact Clarence's alien and have a showdown. This was just after I put the computer on the market."

"Tom, Clarence's alien, explained that our alien, Joe had implanted the idea for the computer into my mind, and that the design had a flaw that could be used to trigger the destruction of the computer and the user at the time of Joe's choosing. He did the mind thing and showed me where the flaw was. That is why we did the recall on the computers we had already sold. I modified the design for future sales, and to replace the ones we had already sold, but then we got together and became paranoid about Tom. It could be that he had used that ploy to get us to change the design to do exactly what he said Joe was doing."

"Eventually, the five of us decided that Tom was the good guy and Joe the bad. Tom had said that he would help us get rid of Joe once the computer problem was straightened out. Once we got that done, we went back to him and asked him to tell us how we were going to get rid of Joe. He said all we would be able to do is destroy the portal he is using and force him to start over again, but that would stave off whatever he has in mind for quite some time. He then showed us the design of the system that will do what he said. He implanted the design of the system into all of us to make sure we get it right."

"What he gave us was the design for an electromagnetic oscillator that is capable of putting out a controlled EMP signal that will immediately destroy all radios and computers, aircraft electrical components, whatever uses energy of any sort to operate. We have the design and operating instructions, and we have the money to build it. The problem we have is that it needs to be employed to a remote location accessible only from the air. We could also buy our own heavy lift helicopter, but we still have the problem of manpower and security. In exchange for helping us, we will foot the cost of development and turn the system over to you after we do what we have to do with it."

"That is such an incredible tale that a science fiction writer wouldn't even touch it. You're right about the Roswell incident. I have read the accounts and even with the evidence, I had a hard time believing that aliens existed. I suppose you are willing to introduce me to the good one to prove the truthfulness of your story."

"I don't think that would be possible Mr. President. You would not be able to ditch the Service for long enough to do it, and we couldn't have them tagging along to pinpoint the location."

"What, exactly, is it that you want me to do?"

"Assign some troops to the project, and a heavy lift helicopter somewhere in the future. We may need access to a top line physics laboratory as well. If you want to document the work for future government use, then the people to do that as well. And finally, security at a remote location at which troops will help do the final set-up of the equipment," Eddie told him.

"Would it work if I just assigned a General Officer with a troop compliment and told him to follow your instructions?" the President asked.

"I would think so."

"And no government funds will be required for the development?"

"That is correct. Your only cost will be manpower and equipment, which is under the military budget anyway," Eddie said.

"Even if I disbelieve your story, it would still make sense to humor you to see what comes of the weapon concept you outlined," President Kelly said.

"It's a win, win situation for you. If the project is successful, your reelection is almost assured, and if we are not successful, then who knows what we will face in the future."

"I really would like to meet the alien, to satisfy any lingering doubts. You tell a good story, and the things you four have accomplished in the recent past supports the story. It obviously isn't a ploy to make more money, since you already have more than you could ever need. I will set something up and have someone call you with the details, probably tomorrow," he said.

They all felt a bit better about the situation now. The President probably gave some signal, because the steward came in at that point and said lunch was ready to be served.

As they sat at the table, the President asked additional questions about the aliens. "What do they look like?"

"They look like a holographic projection, with very indistinct features. I suppose you could say they look human like in some ways. The shape of the head is different, and they appear to have three joints in the arms instead of the two we have. Their hands have opposing thumbs as well. We don't know if this is their true form or not, because the figures are sort of opaque, like they are made of smoke which could be dissolved at any time," Clarence supplied.

"Why do you think they chose you in particular?"

"I don't know that they have many choices. The portals are apparently all at remote locations, probably to assure that whatever keeps the portal open is not interfered with by any machines we use. So if someone doesn't happen along for them to contact, then they simply have to wait until that happens. I suppose they could change locations of the portals frequently, but Tom gave the impression that it is not easy to do whatever is necessary," Eddie said.

"And did they tell you how long they had been here?"

"Both mentioned ten thousand years," Beth said. "They also explained that time was not the same with them, since they are made up of different elements. They require no nourishment because they get it from the natural elements. They also mentioned that they can be regenerated when they are outmoded so to speak."

"What's their purpose for being here?" the President asked.

"I think the essence is to assist the human race to develop, but in such a manner that it cannot be attributed to outside influences. According to both of them, they are not allowed to interfere directly with humans, and when we are outside the radius of their portals, they have no way to know what we are doing. There are more than the two, though we didn't discuss numbers. Each keeps track of what they do and how it works out. They send reports back to the home planet and at the end of their tour of duty, or vigil if you prefer, they are graded on the success or failure of their efforts. It's sort of like a game to them," Bill told him.

"If that's the case, then why is the one trying to destroy the planet?"

"We found out that it would not be total destruction, but enough of the population would be killed off that recovery would be difficult, especially in the technological areas. Tom surmised that Joe was doing it just to judge how long it took the population to recover after the calamity. Kind of like an experiment I guess," Eddie said.

"I can see why you guys had trouble dealing with this. They are apparently advanced so far beyond us that they could just use some weapon and destroy the planet easily."

"They probably could, but their ships, or space craft, are located light years out in the solar system. Tom said and I think I can quote him pretty directly on this, 'our space craft are far enough away that you could not reach them in your lifetime with your current technology'. I got the impression that their mode of travel is something like the old Star Trek movie sets used. They somehow are transported from one place to another almost instantaneously."

"I guess we could talk about this all afternoon. I am sorry if I act like a kid in a candy store, but the concept is so fascinating that I have hundreds of questions, most of which you probably couldn't answer anyway. After you finish whatever you have to do, remember I want to meet Tom," the President said as he got up from the table.

The guests arose too, and they all returned to the sitting room.

"I don't want to keep you here all day. I know you have a lot to do, so I am going to give you my private number, actually, my cell phone, and you can reach me at any time on that if something comes up. Usually, if I am in an important meeting, I pass the phone to one of my guys, so if a stranger answers just ask when you can call back to reach me."

They said goodbye and boarded the helicopter for the return flight to Andrews. Clarence said, "Since I no longer have to stay here, why don't I go back with you guys. I can touch base with the security people and look at the place in Carson City."

"You can also oversee the security arrangements for the property. With the military coming in, we probably have less to worry about, but I would like to have all the electronic sensors installed anyway," Eddie said.

"I can handle all that stuff and you guys can try to get ahead of the game with your production line."

Chapter 22

For the next three weeks Eddie and his friends continued the extra four hour shifts three days a week. Most of the original headpieces had been returned and replacements shipped to the purchasers. They had less than a hundred to locate, and it appeared that most of them were in the military pipeline someplace. They were now ahead of the pace at which they filled orders and had enough of the units on hand to provide a cushion in their development. They could always accelerate the order processing, but saw no need to do that.

Their total worth was now above three billion dollars, though none was tracking it. Eddie had selected a security guy to design the electronic sensors in Carson City, and working with Clarence, everything had gone smoothly. The army General the President had assigned to them was with an airborne unit, probably because Eddie had mentioned the helicopter lift capacity he would need.

The General flew to Albuquerque to meet with Eddie and his partners. They showed him the computer production facility, and explained the security arrangement they had in place. He was duly impressed and commented on the arrangement. "The reason I am showing you this is because we want the same set-up in Carson City before we start work. I have sent a man up to take care of the electronics, but your people will have to decide about the physical aspects."

"How many men do you think will be required?"

"To draw a parallel, we have almost one hundred security people on the payroll here, working around the clock. This facilities overall area is larger than the one in Carson City, but that one presents more of a security challenge. I will leave that to you. During the development process, the security people, plus whoever you want to birddog the design are all that is required.

When we get closer to being operational, more people and heavy lift helicopters will be needed. As we get farther along with the design, you can calculate the size and weight of the components. We are having the casings built off site and delivered in halves. We will take one into the work area to use as a prototype to make sure everything fits where it is supposed to. How much of the internal components we install is going to depend on how much weight the helicopter can lift."

"That's a whole bunch to digest at one sitting. How about I just concentrate on the security arrangement for now, and when that is in place, start worrying about the next phase?"

Eddie laughed. "I am almost as overwhelmed as you with the system design. I know I don't need to tell you this, but I will anyway. Though this project has no classification assigned by the military or government, we consider it to be top secret, and would appreciate it if your troops would do the same."

"I appreciate you telling me that. It helps to be working with people who understand the need for security and take it seriously."

"Have you had an opportunity to use one of our computers yet?"

"No, but I have heard rave reviews," the General said.

Eddie took him to an office where one was set up and motioned for him to sit down at the computer desk. He put the headpiece on him and turned the computer on. He offered no instructions or explanation, simply watched the General's expression.

"Where's the keyboard?" he asked.

"Your mind is the keyboard. Just keep your eyes on the screen."

The computer cycled though the frequencies and got the match. The screen immediately filled with an image, which changed rather quickly. As Eddie watched, the images cycled through a lot faster than what he witnessed with most people who tried it for the first time.

After a few minutes Eddie turned the computer off and addressed the General. "It appears the President, or the military is trying to sell me a pig in a poke," he said.

"What does that mean?" asked the General.

"It means that you are not what is supposed to be wrapped in the package you come in."

"I'm afraid I need an explanation of that as well," he said.

"You might be a General officer, but you are not a ground pounder, or even a paratrooper. I would guess that you are from one of the technology branches, probably a physicist or brain trust of some other discipline."

"Why would you doubt my authenticity when the government sent me to you?"

"I don't doubt your authenticity, I just don't believe you are being straightforward with me. I told the President that he could send whomever he desired to document the proceedings, and they would be welcome. Therefore, why is it necessary to send you masquerading as someone you are not?"

"I am qualified to do everything you have asked us to do, so what is the problem?"

"The problem is that I don't know if the President is the one who doesn't trust me, or if it's the army. I suppose the best way to find out is to ask the President." He took out his cell phone and dialed the number the President had given him.

The phone was answered right away by the President himself. "Did I catch you at a bad time, or can you answer a question for me rather quickly?"

The President said, "Ask away."

"The General they sent to interface with me is not what he appears to be. I wonder if you instructed the military to do this, or if it was an independent action on their part."

"What specifically is the problem?"

"The General they sent to handle security and the troop's supervision has an IQ off the charts, and I don't think he was sent to look after security arrangements. I don't care if he stays as an observer, but I hate to have them think they need to pull something like that when we agreed that you could have as many people as you wanted doing the documentation."

"Is the General there?"

"Yes sir, he is."

"Does he know who you are talking to?"

"He probably thinks I am trying some scam on him, so I would say no."

"Put him on the phone."

Eddie handed the phone to the General, who reluctantly took it. "This is General Langdolf," he said.

"This is the President. I want to give you some advice General. The people you are dealing with are smarter than you by a magnitude of hundreds. If you were told to spy on them, your mission is unnecessary. They will give you open access to all they do and all you have to do is take notes. I think what Eddie is worried about is that you will look at your spy mission as paramount and neglect a very important aspect of the operation. Do not try to deceive this group. They will throw you out on your ear and probably withdraw the agreement to give us the hardware when their use of it is over. Is all that clear to you?"

"How do I know that this is the President?"

The President made a quick decision. "You will know it in about thirty minutes when the Chairman of the Joint Chiefs makes the phone call to relieve you. Your services are no longer required. Now put Eddie back on the line."

He handed the phone to Eddie. "Yes sir," he said.

"I agree with your assessment. I will have Langdolf relieved, and make sure my instructions to the military are a bit clearer this time. I apologize for the snafu. Call if you need anything else."

Eddie hung up the phone and turned back to the General. "I don't know if this was your idea, or whoever sent you, but the President's intention was not clearly understood by someone. You can wait outside until the call comes for your relief."

"You're joking. That wasn't really the President. People don't just call him on a cell phone?"

"Are you a betting man General?"

"Why do you ask?"

"I will give you fifty to one odds that the Chairman of the Joint Chiefs of Staff calls this office inside of thirty minutes to talk to you," Eddie told him.

"And what will he say?"

"Words to the effect of 'pack your bags and go back home', or, 'whose hair brained idea was this'? Either way, you are not going to be part of this project. He escorted the General to the waiting room and just as he was turning to go back to his office, his cell phone chirped. He looked at the caller ID and handed the phone to the General, who looked and started to hand it back.

"You might as well answer it, it's for you anyway."

It was indeed for the General, and the conversation was a short one.

When he hung up the phone and handed it back to Eddie he turned and walked out of the building. The phone rang again within thirty seconds. It was the Chairman, and he asked, "Is the asshole gone?"

Eddie chuckled, "Yes sir. He didn't utter a word, just turned and walked out."

"I apologize for that. The orders I gave apparently were misinterpreted. Can you tell me exactly what you need so that I can be specific when I pass this next order on?"

"I need about a hundred good security troops, and someone to lead them that knows the business, preferably no more senior than a Major. When the units to house the components arrive we are going to put one inside the building and use it as a prototype to make sure everything mates up the way it is supposed to. You can send as many people as you want to do the documentation, but I suggest more than four and they will be getting in the way. They need to be technically oriented, probably theoretical physicists, mathematicians, or electrical engineers with a lot of experience. Later on we will need a heavy lift helicopter, maybe as many as four, and more troops to perform some manual labor getting these things where they have to be."

"Would there be a chance for me to come out and have a look at what you are doing?"

"Certainly, you are welcome at any time, but I suggest you wait until we have all the equipment procured to build the thing before it will be very beneficial to you. Did the President give you any idea what this was all about, I mean the system we are building?"

"No just that it would be beneficial to the military and that it behooved us to be cooperative."

"Come on out whenever you can work it into your schedule and I will give you a brief on the capability you will be getting eventually, at no cost I might add," Eddie said with a chuckle.

"Your new man will be there tomorrow. I think the troops they sent will probably do, but his instructions will be to follow your orders. If the troops are not satisfactory for any reason, let him know and he will take care of it. I am thinking of sending my Aide to you. I will not be able to find a replacement for him who is as good, but your need is greater for what he can offer at this time."

"Thank you for being understanding General. I didn't want this to get off on the wrong foot."

"I will be in touch in the next few days. Where do you want the Major, in Albuquerque or Carson City?"

"Route him through here and I will get him along the way after we talk."

"You've got it. Talk to you later."

Chapter 23

Major Dalton arrived the following morning and Eddie invited him in for a briefing. "I assume the Chairman told you as much as he knew, which was not a lot. We have a facility in Carson City, Nevada where we are going to build what we call an electromagnetic oscillator. It is light years ahead of anything in our knowledge base. We are building it for personal reasons, but when we are finished with a onetime use, we will turn it over to the government for possible military use. I can't give you a good description of its capabilities because I don't know at this point what they will be. It can be used to generate electricity at a very economical trade off, or as an electromagnetic pulse generator for offensive warfare. It will have the capability to render everything within its range of influence inoperable. I am talking electrical, mechanical, computers, airplane engines, including ballistic missiles, and probably some other things I don't know about yet."

"What's my role in this?"

"You are going to command the troops and be responsible for the security of the plant in Carson City, Nevada. Size wise it is smaller than this facility, but might be more of a challenge because it doesn't have the perimeter obstacles we have here. I want you to treat the project as top secret, with strictly controlled access, escorts for visitors and delivery people, and a ready reaction force. We use about a hundred people here, but you can structure the Carson City plant any way you want that will be effective. I am not so much worried about other countries, as I am about the threat from within the United States. Word will leak about this, but not the particulars hopefully, and people will be curious enough to try to find out what is going on. Your job for now is to see that doesn't happen. Impress the troops of the gravity of the situation. For reasons that I cannot reveal, there is the outside chance of a concerted effort to breech the facility by some very serious and dangerous people. I can't give you any more than that, just that the possibility exists."

"In other words, you want the best security that we can provide, against the possibility of a concerted effort to breech the place."

"That's it in a nutshell. It probably will not prove to be needed, but better safe than sorry."

"My General said to tell you that he would try to get out to Albuquerque the first part of next week. He will call you with details. How am I going to get to Carson City, and who will be the contact point there?"

"Clarence Woodman, who you have probably heard of if you follow the financial world at all, will be on site. He is in his seventies, but don't let that fool you; he is very sharp, and probably can take any of your troops one on one if he chose to do so. His money is going into this project along with ours, and he has a personal stake as well. When we get further along, I will let you know what additional requirements we will have. I expect that we will have to use a physics laboratory belonging to some other organization, but until we start the drawings, we will not know. You can find out which helicopter the military uses with the highest lift capacity. We will definitely need one of those, possibly as many as four, but they won't be needed until well along in the project."

"Can you give me any sort of timeline?"

"I'm sorry, but we just don't know yet. It could be as little as three months or as much as a year?"

"Just so I understand, my job is security?"

"That is your primary focus at the present time, but as we need other things, you will be the person we come to for the coordination efforts with the military. I have a man on site now looking at the sensor set up. You will want to get with him to make sure everything becomes operational as soon as possible. His name is Kinsey, and I pulled him off the job here to do that."

"Okay, how do I get to Carson City?"

"A business jet, belonging to Woodman is available when needed. It is at the airport here now, so I will call the pilot and tell him to expect you and to get you to Carson City. He will probably fly into Reno, but it is a short hop on down. I am thinking about buying a helicopter for that sole purpose."

"We will need cars, probably a good many, along with some military vehicles. You can work that out, and also the armament you feel necessary for the job. I will arrange to have some cars delivered tomorrow, probably from a local dealership. Let me get someone to take you to the airport and you can be on your way."

Eddie called security and asked that someone take the Major to the airport, then he started on his to-do list. Since the last thing he thought about was the cars, he called Clarence and suggested that they lease a dozen cars from a local dealership and have them delivered to the work site. He then suggested that they lease a helicopter to ferry people and supplies from Reno to Carson City, since they would be using Reno as the hub for the business jet.

Major Dalton arrived at the Carson City facility just before four in the afternoon. The General he relieved had not done anything about housing and other orders for the troops and he found one hundred people scattered around the compound. The Captains and Lieutenants had organized them into squads and had a perimeter security of sorts set up. He called them all together inside the hangar like structure and told them to wait for him there. He located Woodman and introduced himself and told him he would be in charge of the security detail and would do the liaison with other military organizations as the need arose. "I am going to get the troops organized and set up a watch rotation. As soon as everything is organized I will get it to you for approval. What are we going to do about office space?"

"How much will you need?" Clarence asked.

"Probably a couple of hundred square feet. Ideally it will be where the electronic devices are monitored."

"Why don't you pick a space along the perimeter of the building and draw out what you want and we will have it built. Coordinate with the security guy Eddie sent up and tell him how you want the monitors located. He might even be grateful for some help in designing the surveillance fields of the different components."

"Point me in his direction and I will do that," Dalton said.

Clarence found the man and introduced them. "The Major's people will be handling the security when everything is ready, so make sure you set it up the way he wants it."

After Clarence left, the man told Major Dalton that he was trying to duplicate the way they had the other facility set up, and that he was not a security specialist, so any input would be appreciated.

Major Dalton took the man with him back to the area where the troops were waiting. "Who's the senior officer present?" he asked.

After glancing at each other a Captain raised his hand. "I believe I am sir," he said.

"I want you to divide the men into squads of ten and set up a rotation in four hour shifts to cover the remainder of the day and night. Tomorrow we will have things a little better organized and I will brief everyone on the mission and our part in it. Treat this facility like you would a military top secret facility. Nobody comes or goes without an escort unless they are on the access list. That will be provided tomorrow. In the meantime, anyone wishing access will have to be approved by myself or Mr. Woodman. No exceptions! Does everyone understand that?"

"Sir, where are we supposed to sleep?" one of the Corporals asked.

"What did you bring in the way of equipment?" the Major asked.

"Just our personal gear sir. This was laid on kind of quick and nobody knew anything back at Bragg."

"Once the watch rotation is set for the night, those not on the list can go and find hotel rooms. Those with the watch can sack out here wherever you can find a soft spot. I will address that issue tomorrow as well."

"I want the rest of the officers and non-comms to come with me."

He turned and motioned for the security guy to follow him and led the group to the office portion of the building. Mr. -" he paused for the man to give his name - Hamlet he supplied, "will be installing sensors starting tomorrow. We need to help him decide what we need. Motion sensors, video cameras, infra-red and sound pickups will all be included. I want you to show him where to locate them, and how to code them so we know if an alarm sounds where the breeched point is located. I want everything to terminate in a command center inside the building. Choose a location out of the way, but within easy access to doors for a reaction force. Are there any questions?"

"Sir, this isn't a drill is it?" asked one of the Sergeants.

"No drill, Sergeant. We are going to be here for a long time, and this project is being directed by the Chairman of the Joint Chiefs of Staff and comes directly from the President. We will have additional duties as time goes by, but I can't tell you what those are at the moment. The morning briefing will be a lot more comprehensive and should help you understand the secrecy a bit better."

"Captain, see if you can locate Mr. Woodman and tell him we need a paper layout of the property. Maybe you should ask for a dozen, so we won't all have to work from the same page. When the troops do their rounds tonight, check the condition of the fences. If there are breaks or damage to any sections, let me know and they will be replaced or repaired tomorrow."

"Okay, everyone have a look at the property outside while there is still light, with a mind toward where we want sensors and what type."

They went their separate ways to get on with the work. Major Dalton had a tour of the outside on his own. He was primarily looking for locations to place weapons in order to repel an attack. The fence was a good distance from the building, but the building was not all that substantial. A grenade or mortar round would breach the thin metal with little trouble. It would be easier to sandbag the inside of the walls than to try to strengthen them from outside. The prime objective was to protect whatever was going on in the working space, so a six to eight feet layer of sandbags would provide all the protection they needed against small arms fire and grenades. It might tear up the building, but hopefully wouldn't do a lot of personnel damage. He made a note to look for sandbags the following day.

The roof was metal and sloped on two sides. It would not be possible to mount an observation post on the roof, but he could use a scissors lift in the corner and cut out an observation/firing port. He made more notes. As it started to get dark he went back inside to evaluate what his men had come up with for sensor locations.

These people were all seasoned veterans and looked at the defensibility of the place as well as the detection capability they needed.

Between them, they had the entire compound covered. Once they combined the recommendations into a single document, they had it all covered. The IR sensors would be nearest the fence, with motion detectors inside those, but nearer the building. Cameras would be mounted on each corner of the building and in the center of each side. A guard house would be built, not very pretty, but effective against anything short of a tank, with heavy duty traffic bars. This could all be done relatively quickly and bulletproof glass would be built into the design

The Major incorporated his thoughts about the sandbags for the inside and the idea for scissor lifts for observation platforms. They would look at the final product in the morning and get it to Woodman for approval.

Eddie had given Major Dalton a cell phone with pre programmed numbers for the four of them as well as Clarence. Dalton now called Clarence, who had gone to his hotel. "What are we going to do about accommodations for the troops?"

"I guess I just thought you people would take care of that. Never assume anything huh?" he said with a laugh.

"I have a watch rotation set up, but I still have about fifty troops with no place to sleep."

"Have them try to find hotel or motel rooms for tonight. I will address the issue with Eddie and come up with something tomorrow."

He called Eddie right away. "We neglected to consider the logistics for the people. Where are we going to put them up?"

"What would you think about buying a hotel?"

"Are you serious?"

"Why not. If we can find something with fifty rooms, we can double them up and after this is over, either sell the hotel or hire a management company to run it."

"Isn't it nice to have enough money to make a decision like that off the cuff?" Clarence asked.

"It sure is. I never dreamed that I would be in a position to be able to spend three or four million dollars and not give it a second thought. The bottom line is that we need it, so what do you think?"

"I guess it is the easiest solution, and probably the best one as well. I will start looking in the morning. In the meantime we will reimburse the troops for their expenses until we get something workable."

"I take it Major Dalton is working out all right?"

"He took charge right away and the troops respond well to him. I think he is the right man for the job."

"Talk to you tomorrow," Eddie said as he hung up the phone.

Chapter 24

On the following morning Major Dalton got together with the security man Eddie had sent and they went over the work the Major's people had done the night before. "I think you can go ahead with the sensors. I suppose you are going to contract that?"

"It wasn't discussed, but the job is certainly beyond my capabilities," he replied.

"Let me call my office and get the names of some of our better contractors for that type of thing and maybe you can start there."

"I will clean up this copy of the plan so we can use it for the contractor to work from. It isn't all that elaborate, but I think it will probably suffice for the job."

"The stuff inside the building I will get my people involved with. I'm talking about the sandbags and observation posts, not the wiring," Dalton said.

Clarence came in while they were talking and Major Dalton asked if he had a few minutes.

He showed him the rough plan and asked if they could start looking for contractors.

"We could probably do it on our own, but the contractors can do it faster, and everyone else is very busy."

The security man asked about authority to spend money and Clarence told him to bill everything to the company. "Major, how about having your troops do the trenching for the lines for the sensors. You seem to have a good handle on the way you want it run, so that would help out a lot. I can rent trenchers or whatever else you need."

"Yes sir, we can certainly do that. I want to get enough sandbags to line the interior of the building up about six feet as protection against small caliber projectiles if it comes to that. The metal building doesn't provide any protection at all. I also want to purchase four scissor lifts to create viewing ports in the upper corners of the building. You can think of them as watchtowers at a prison, but with a restricted view and line of fire."

"You really are preparing for a worst case situation, aren't you?" Clarence asked.

"Yes sir. That was the orders I got from Eddie. I don't know enough about the situation to make judgments on my own, so whatever you guys tell me, I take as gospel, and Eddie said prepare for a worst case scenario."

"I hope it doesn't come to that, but you are right of course."

"I also need to brief the troops a little more fully. Maybe you and I can sit down and talk awhile before I do that."

"I talked with Eddie last night about housing for the troops. He wants to buy a hotel large enough to house them all, if we can find one to meet our needs. I have someone looking now."

"You're going to buy a hotel just to put the troops up?"

"Unless you can come up with something more workable."

"We could probably erect temporary military style tents inside the property that would suffice."

"The drawback I see with that is that it will draw additional attention to the area, and along those lines, you might want to have the troops wear civilian clothing, maybe security uniforms, which I think we can find easily enough. I think the hotel is the better option and Eddie pointed out that we can hire a management company to run it after we finish with it."

"I agree with those points. We also need to think of a way to feed everyone. If there is room in the hangar, I can have a military kitchen set up to provide food."

"That's a very good idea. We will supply the food if you can arrange that. What else have we not thought of?"

"I'm sure there's something, but we have enough to keep busy for a good while. I want to brief the troops to impress them with the need for vigilance and the importance of what we are doing. I can be general enough to stress the importance without getting into details. Is there anything you want to tell them specifically?"

"I will sit in on your brief if you don't mind and answer any questions I can."

"I would appreciate that."

Dalton had the Captain gather the troops in the hangar and he addressed them. "First, we are not going to be wearing uniforms for this assignment. All of you need to give your shirt and trouser sizes to the exec so we can get civilian security uniforms ordered. Four sets per man. You will continue to look out for your own lodging until other arrangements can be made. Mr. Woodman is negotiating for a hotel to put everyone up in the same location. That might take a while, so if anyone is short of money for living expenses, let me know. You will be reimbursed for your expenses, so keep receipts. I am going to look into setting up a military kitchen in the hangar so we can take our meals here."

He turned to the Captain, who by default was the executive officer. "Look into a mobile kitchen. Find out where the closest one is and we will get the movement orders issued through JCS."

"Now, the reason we are here is to provide security for the development of a new weapons system. I don't know much about it, except that the President thinks it is worth our efforts. The work will be done by five civilians, including Mr. Woodman here. They are footing all the bills for the development, and the project is top secret. Once we get into the development stage, the facility will be treated just like a military top secret facility. No one will be granted access to the site without approval of one of the five people I just mentioned unless they are escorted. All deliveries will be witnessed by one of you. The block house for controlling access should be done in two to three weeks. We don't anticipate any work beginning for that long, so everything should be in place by the time they start."

"The nature of the project does not deal with munitions or anything that could accidently go boom, so rest your minds on that score. On the other hand, things don't have to go boom to be lethal."

This got a chuckle from many of the group.

"I know that our method of operation is unorthodox, but that can't be helped. The importance of this project is right up there with the Manhattan Project during World War II if you young folks can relate to that. That was the code name for the effort to develop the atomic bomb in case some of you don't know."

"We don't know of a specific threat to security, but I have been told to prepare for a worst case scenario, so don't take this as a peachy assignment and neglect your duties. Once the hardware is done the job gets exciting. We will have to transport the completed unit to an as yet unknown location for its maiden use. The system will consist of four units, about the size of a small bedroom. You will get a better feel for the size as the work goes along. That's about it. Mr. Woodman will take your questions now. Oh, one other thing before I turn it over to Mr. Woodman. If any of you have experience with trenchers, back hoes, or such, let me know. We are going to do the trenching for the sensors before the contractor gets here to install them. Now, questions?"

"Sir," a Sergeant Major said, "what about weapons and rules of engagement?"

"I'm glad you brought that up Sergeant Major. You will be armed with side arms and semi automatic rifles. I haven't decided how to set up internal defenses yet, but I am thinking at least two machine gun emplacements maybe set up inside the building with a door that can be opened to reveal a line of fire. If we can get away with it, I may even ask for a couple of tanks."

Some more chuckles could be heard.

"I am not joking gentlemen. We have no idea about any force that might be brought against us, and I want to err on the side of caution."

Nobody had any additional questions and the meeting broke up.

Later in the morning low bed trucks brought two trenching units and a back hoe and unloaded them. The army troops who indicated they knew how to operate the equipment went to work. The trench lines had already been marked with spray paint and the machines were busy for the remainder of the day.

Several places in the fence needed to be repaired or reinforced and the material ordered for that chore. The troops seemed to be happy to be occupied and they even dug the foundation trenches for the block house.

There was enough expertise in the unit to do the majority of the work, and instead of looking for a contractor to build the blockhouse, they decided to do it themselves. It is amazing what a hundred people can accomplish in a short time.

The same company who had put in the security system in Albuquerque was doing the work in Carson City and they were on site the following day. Since the trenching had already been done, they simply installed the sensors and ran the lines in the trenches provided. As the line was laid the army troops filled in the trenches and all the outside work was completed in two days. The console for the security system was rack mounted, and it was simply a matter of installing the equipment in the racks and hooking up the lines. They had to test it and that took longer than the installation. Within a week all was ready, except the block house, and it took four days to get the bulletproof glass, which held them up on that item.

A hotel had been found that would accommodate all the troops, and like the hangar building, they offered a price above market value and closed the deal within two weeks.

The first half of a housing unit was delivered by flatbed truck and it was taken into the hangar and positioned where it would be available for the work. More machinery arrived and was set up on work benches as Clarence directed.

Raw materials started to arrive. The first item was rolls of copper wire. The transport vehicles could only carry three of the rolls and they had ordered twenty four rolls, so it took eight loads to deliver it all. That happened over a ten day period.

Other equipment arrived piecemeal. Grinders, lathes, drill presses; all manner of tools were delivered at all hours of the day. If Clarence was not sure about a tool, he bought it anyway. They would need some precision instruments, but he wanted Eddie to handle that part of it. In a phone call, he told him as much, and Eddie did the same as Clarence. If there was doubt, buy the machine.

The hangar began to fill up the available space, and the actual work had not even started yet.

It was almost three months before they felt they were ready to go to work. In the interim, the security troops had used metal strips to interlace in the chain link fence to hide the activity inside the compound from view. It didn't shield it completely, but one would have to be right up to the fence to see what was taking place inside the compound.

The Chairman of the Joint Chiefs of Staff had visited Eddie in Albuquerque and Eddie had briefed him about what they were building. He showed him the design of the system that he had printed out for working documents, and though the General was not a very technically educated individual, he understood the concept of the design.

He commented, "This looks like something from outer space."

Eddie though the comment was just coincidental and let it slide. He didn't believe the President would even confide to the General about where the design really came from.

"It is some really advanced technology and concepts. We are not sure about the output of the working unit, but we should be able to get some measurements as we go along. The basic concept is to have the output from four units intersect at a given point, which will generate the EMP upward in a cone pattern. Everything within that cone will be obliterated, electronically speaking, based on the power output."

Eddie continued, "As soon as everything is in place in Carson City, we will go to work. Our computer operation is well ahead of schedule now, and two people can keep up with the programming to make deliveries on time. Do you realize that we will exceed three billion in sales next week?"

"That is simply amazing, but the computer is amazing too. I know the military wants to buy thousands of them, and I imagine industry is clamoring just as bad as we are."

"The potential for sales has been calculated at over five trillion dollars. Can you believe that?"

"What I can't believe is that you are selling them so cheap, relatively speaking. You could have sold them for a million dollars each and nobody would have balked. The technology is that advanced."

"We didn't want to price the average guys out of the market. They deserve them as much as the wealthy we felt and priced them accordingly. We are still making a lot of money, but at some point money is no longer the important thing."

"What is the real objective of the project you will be working on with the EMP device?"

225

"I'm sorry, I can't tell you that General. We even had trouble convincing ourselves to confide in the President. We are going to have a onetime use for it, but with the potential we felt the government would make use of it, at the very least to generate electricity very cheaply. I did mention that aspect of it didn't I?"

"I don't believe you did. You are saying that the device will double as a generator for electrical power?"

"Yes sir. It will do that much more cheaply than any method we use currently, including nuclear."

"What's the output going to be in megawatts?"

"We're not sure yet, but enough to power a small city at least."

The General whistled. "You could have marketed this to the power companies and become a nation unto yourselves."

"Like I said, it isn't about money. We have a job to do that requires the unit. None exists, so we have to build our own. You need to think about who you will use to do the documentation. You can see from the design plans that they are going to have to be very smart to keep up with us. If I gave you the design plans now, I don't think anyone would be able to interpret them to the point that they could build one of the units. Someone is going to have to observe our methods to be sure you have enough understanding to use the plans later if you decide to do so."

"I appreciate you taking the time to explain this to me."

"You want to come to Carson City about a month into the project. You will get a better appreciation for the size and scope of the physical unit then. By the way, thank you for Major Dalton. He is every bit as efficient as you said he would be, and isn't afraid to make decisions."

"He may very well be sitting in my chair one of these days."

"You could certainly do worse," Eddie said.

"Is that an oblique reference to General Langdolf?"

"Not intentionally, but it could be," Eddie laughed.

The General left with the feeling that he had been in the presence of a true genius, and any misgivings he had about the military involvement were dispelled.

Though Eddie didn't know it, other sinister forces were at work. The facility in New Mexico was under surveillance again, this time by characters more unsavory than the original group.

They were also more circumspect, and more numerous. The leader behind this group was a Mexican drug lord, who had happened upon one of the alien portals in south Texas a few months earlier. He had been approached pretty much as Eddie and his group, and Clarence some forty years earlier.

The difference was that the drug dealer had been in frequent contact with his alien, who we will call Harry. He had done the intelligence boosting and given the man ideas about how to make money very quickly. The drug dealer was not too smart to begin with, and his brain had been affected by his product at an earlier age, so there were limits to his creative effectiveness.

Harry wanted to know what was going on in the earthly confines and set Carlos, the drug dealer, to work finding out. He told him what to look for and when Carlos told him about the new computer coming on the market, he coaxed him to try one out, which he stole from one of the first buyers.

At their next meeting not a week later, he probed Carlos's brain and saw that the design of the computer was based on the headpiece they used. He extrapolated that the only reason this could be was that some of his brethren were not playing by the rules.

He had put Carlos onto ways to make money that he had never thought of, all legitimate, and inside three months Carlos was a multi-millionaire. He then suggested that he start to build a paramilitary group on the chance that they might be needed.

When Carlos got wind of the project in Carson City, it was suggested that he place a watch on the facility, and on the computer facility in Albuquerque as well.

With Carlos's connections he was able to amass a rather large force and chose the smarter ones to lead the groups with the order to stay out of sight, and report to him by phone daily.

Harry was sure that one of his kind had been responsible for the development of the computer. The news of the facility at Carson City worried him that a war was about to erupt between two or more of his kind. He would know more as Carlos gathered additional information about the new facility. He suspected that there was no way the earthlings could harm his kind, but he was not sure. If one of his brethren knew something he didn't, then this could be a ploy on their part, and he could not take that chance.

He very subtly guided Carlos in more money making schemes so that he would be able to fund the effort. Carlos's increased intelligence had one positive effect that Harry was thankful for. It had shown him the travesty of drug use and made him see the self destructive aspects of that habit. He has slowly withdrawn from the habit and for the last month had not touched the products he used to peddle.

Harry felt he was at least a bit more reliable now.

He would just have to wait and see what developed.

Chapter 25

Eddie had about decided that he and Bill could head for Carson City the next week. He was sitting at his desk going over the things he needed to do before they left, when there was a knock on his door.

The security chief came in and Eddie offered him a seat. He obviously had something on his mind and Eddie waited until he could get his thoughts straight. He finally said, "I don't know if it is worth bothering you with, but something funny is going on."

"Funny how?" Eddie asked with a smile.

"I don't know how to explain it, but things are just not normal."

"Okay, what is not normal?"

"I send guys out a couple of times a day to drive around the area and check things out. They have noticed drug dealers on street corners where they shouldn't be, and other characters that don't seem to belong to the neighborhood. It isn't anything they have done wrong, just that they don't fit in with the landscape."

Eddie's antennae went up. "Give me a little more," he said.

"The drug dealers don't seem to be selling much, and they have been seen talking to some of the other people who don't seem to belong to the area. The numbers are large enough to be disturbing if they have designs on this facility and that worries me."

"You think they might be keeping an eye on us here?"

"Yes sir I do, but I don't know why, or who they would be working for. They are mostly low life types, people you would expect to see in the night club section, or at the casinos. They just don't blend in."

"What kind of numbers are we talking about?"

"I can only give you what my people have told me, but that number would be between twenty and thirty, and there are probably ones they didn't recognize, or in cars they didn't identify."

"If a force that large attacked us here could we hold them off?"

"It depends on how they are armed, but we are not structured to handle something like that."

"Excuse me a moment while I make a phone call."

The security chief started to rise to leave the room, but Eddie motioned him back to his seat.

Eddie called Clarence. "I am here with my security chief, and he tells me we are being watched by twenty to thirty unsavory characters. I wonder if the same situation exists there."

"I don't know, but I will put a bug in Major Dalton's ear to find out. Can you tell me anymore?"

"They seem to be drug peddlers, or other unsavory types of the same ilk. He suggests that they would be more at home in the bar or casino districts."

"Do you have any idea what this might mean?"

"It could be that there's another player in the game, or Tom is playing both ends against the middle."

"I will call you back when Dalton can find out something."

Eddie then made another call to General Clarkson, the CJCS. The call was answered by a secretary. "General Clarkson please, Eddie Casteen calling."

The General came on the line right away. "What can I do for you Eddie?"

"Funny you should ask. Can you spare another fifty troops with heavy weapons?"

"What's this about?" he asked.

"I just learned that there are twenty to thirty, maybe more, people watching the facility in Albuquerque. I suspect the same is going to be true in Carson City. I gave them a heads up a few minutes ago and Major Dalton is on that. I am worried about this facility because all we have is the civilian guard force, and they are not set up to repel a concerted attack by that large a group, especially if they are heavily armed."

"Do you expect anything imminent to happen, like today or tomorrow?"

"I don't know. This just came to my attention, and the watchers are posing as drug dealers and street vagrants. I don't know how long they have been in place, nor what their intentions are. They don't appear to be set up for an attack in the immediate future," he said raising his eyebrows in question to his security chief.

"I can get some troops on site in civilian clothing, but it is going to be difficult to get heavy weapons there right away. We will probably want to send them in a semi as a regular delivery. How many troops do you need?"

"Special Forces or equivalent, maybe twenty. We can use my security people to augment them as necessary."

"You have no idea what this is all about?"

"Yes sir, I have an idea, and if it plays out that way, these guys will be well funded, and well armed."

"Are they a threat to the country, or just to your facility?"

"You could put them in the same category as the Carson City facility. If you want, I can ask the President to okay this to keep you from getting into a bind."

"If I sent you twenty combat veterans on leave from their stations could you employ them as part time security troops?"

"Absolutely. I suppose you want one of your armories to be burgled and some heavy weapons stolen?"

"Not a bad idea. Do you know of one close by?"

"You probably know more about that than I do."

"Let me check into it. I will get some people moving first, so it might take me a while to do this. If you think a call to the President is in order, then make that call while I do the other stuff."

Eddie hung up and placed the call to the President. "I hate to bother you sir, but I just got off the phone with General Clarkson and he is providing some support for me at the Albuquerque facility. I think it involves the same people we talked about at Camp David, but a third party."

"Do you mean another Tom or Joe?"

"Yes sir, it appears that way. I think they have the Carson City facility under surveillance as well. I am not as worried about that since we have a hundred troops there, but Albuquerque is vulnerable. General Clarkson is setting something up that will hopefully keep the government out of the mix."

"Is this something he is comfortable with?"

"It seems to be. He is going to call me back in a few minutes. If you wish I will ask him to brief you fully after we get the action started."

"Please do that. I take it he is still in the dark about the real reason behind all this?"

"Yes sir, but he is very curious?"

The President laughed. "I can just bet he is. If this is a third party as you suspect, what impact will it have on your plans to get rid of the other party?"

"I may have to pay Tom a visit to get some guidance. It would be a good opportunity for you to meet him if something could be arranged."

"Let me give it some thought. Have Clarkson come see me after you get everything in place."

"Yes sir, will do."

After he hung up he turned to the security chief. "Did you follow that?"

"If I heard right, you were asking the Chairman of the JCS to send some troops in civilian clothes, and some heavy weapons in a semi. You then told the President that the Chairman would brief him," he said in awe.

"You got it right. Now those conversations, and my involvement with the people in particular is classified, so keep that to yourself. You needed to know because you will be doing the coordination with the troops when they show up, which will probably be first thing in the morning, or even later tonight. I don't want you to send anyone out alone from here on out. I suspect that this is going to get interesting soon, but I don't know how soon."

Eddie's phone rang. It was Dalton in Carson City. "I do believe we have a threat. I haven't pinpointed the size of the force yet, but I am sending reconnaissance troops out to assess the threat."

"Your boss is sending some people to Albuquerque. They are military people on leave, who will be added to our payroll as temporary employees. An armory at some unknown location will be burgled and some heavy weapons taken, but it will not be reported for a few days. How are you fixed for armament?"

"We're okay. I have trained troops, and unless the opposition sends tanks, we can handle them. Does this have to do with the threat you expected to face?"

"I don't know yet. It might be a similar threat, but unrelated. I am going to have to make a trip back east soon, probably as early as tomorrow, but I will have my sat phone, so call if you need anything."

He then called Clarence again. "I am going back east, probably tomorrow to see Tom. Something about the latest doesn't add up, and maybe he can shed some light on the situation. I suggested to the President that it might be a good time to meet Tom if he can get away. What do you think about that idea?"

"It's a good one. Who's going with you?"

"Hopefully you are."

"Where's the plane now, here or there."

"I think it's in Reno. You can swing by here and pick me up. Dalton will be able to handle things there."

"Do the others know about this yet?"

"I am about to tell them right now."

"Okay, talk to you later."

Eddie and the security chief left his office and went their separate ways. Eddie found Cindy and Bill on the production line and told them they needed to talk.

They all went back to Beth's office and Eddie told them what had transpired over the last half hour. "So Clarence is going to swing by here and get me and we are going to Washington and see if the President can break away for a visit to see Tom. Questions or comments?"

Bill asked, "How do you want to handle the situation here?"

"When the troops start to arrive, add them to the security payroll and brief them on the situation. The security chief is due for a bonus, and he can fill them in on the people and where they are. The army should be able to handle it from there. I got the impression that Clarkson was going to send a couple of Special Forces teams, so they can handle themselves."

"Do you really think there is a third alien involved now?"" Beth asked.

"Either that, or someone slipped by us with Joe," Eddie said.

"Tom is not going to know the location of a third party any more than he would know Joe's location," Cindy observed.

"That's probably true, but there's a lot about their existence we don't know, so maybe Tom will be able to put us on the right track, or at least explain what is happening now."

"It's a good thing we designed good security into this place, and Carson City too, for that matter," Bill said.

"If you guys think it is necessary, shut down the production line and send everyone home for a few days. We have enough inventory to keep up with orders for a good while, so nothing would be lost."

"I think we should just keep chugging along until we see how this plays out."

"One thing I think we should do from this point forward, is be armed at all times. From what security says, these guys are several steps down the food chain from the people Clarence had watching us, and they will not be so gentle in a confrontation," Bill said.

"I agree with that whole heartedly," Cindy said as she went to her desk and took out the pistol she always had near since Beth's abduction.

Eddie's phone chirped and he looked at the caller ID. "General Clarkson," he said, and answered it.

"Okay, here's the plan. You will be getting twenty two people over the next twenty four hours, in ones and twos. All will be in civilian clothing and all will have a personal weapon with them. A U-Haul truck will be arriving before noon tomorrow with heavier weapons and the people coming will know how to use them and how to set them up."

"Thanks General. I talked to the President and told him you would be over to brief him more fully on the actions we took. Clarence and I are coming to Washington tomorrow, and may be taking a short trip with the President to Camp David. I will call you when I get there with updates as necessary."

After the call, Eddie said to the others, "What if we have the Special Forces guys abduct one of the watchers and see what information we can get from him?"

"That might tell us a lot. Where the guy is from might be telling in itself. They won't have any trouble getting them to talk. It might be as simple as waiting until they start to suffer withdrawal symptoms."

The others laughed.

"You could be right. These guys are a long way from what I would have expected from someone with our intelligence, but that's only an assumption," Eddie offered.

"The way these portals seem to work, the aliens don't have much choice in the matter; they just have to take whoever happens along their path."

The first two troops showed up later that evening. They had been on a temporary assignment to Fort Huachuca in Arizona and had been told where to report, but nothing additional.

Eddie explained the situation to them and turned them over to the security chief for a briefing on the opposition.

They decided to do a bit of surveillance on their own before the others arrived and would take one of the company cars. Before they left they used a bathroom to change clothes. When they emerged, everyone laughed. They looked like street bums, right down to the reek of stale alcohol.

"We can be more effective this way. Nobody expects a street bum to be anything else, so we can get closer to the druggies and their suppliers than normal before anyone becomes suspicious," one of them said.

"I like it," Cindy told them.

Chapter 26

Clarence called Eddie a little later and told him that Major Dalton had ascertained that a good number of people were watching the Carson City location as well, and they appeared to be the same caliber of people observed in Albuquerque. "I will be leaving shortly and should be there around midnight. Meet me at the airport and we should be in Washington before eight with the time change."

They did as planned and took a cab to the hotel where Clarence usually stayed. Eddie made the call to the President after they had breakfasted and he told them he would have a car pick them up and bring them to his office. That way there would be less chance someone recognized them, both of whom were now well known.

The meeting took place in the Oval Office. Eddie asked, "Did General Clarkson brief you last night?"

"Yes. Is there any additional developments?"

"No sir. They seem to be in a watch mode only right now. Did you give any thought to my proposal to meet Tom?"

"I think there might be a way. I assume the location is not terribly far from here, or from Camp David," he said.

"It would be better if we could get to Philadelphia by air, then the drive would only be about half an hour," Clarence told him.

"I can't ditch my entire detail, but I can reduce it to two, plus you and Clarence, then have them wait in the car with either of you while the other takes me to the meeting. Would that work?"

Clarence and Eddie gave it some thought. Even if they observed the meeting, all they would see would be them standing under the trees, apparently having a private conversation. Clarence said, "I think that would work."

The President called his lead agent in and told him what was going to happen. "I am going with these gentlemen to a location I don't want disclosed to anyone, and I mean absolutely no one. I am going to take only two of you along, and you will have to wait in the car while we have a private conversation. Are you okay with that arrangement?"

"Will there be anyone else close by?" asked the agent.

Clarence handled the reply. "No. It is a rural setting, and a state highway is close by, but nothing else."

"Will we be able to observe?" he wanted to know.

"From a distance."

The agents apparently gave it some thought and the lead agent replied, "If you are okay with these gentlemen, then that will fulfill our requirements."

"Okay," the President said, "have my appointments put on hold until two o'clock. No explanation just that the schedule had to be rearranged. Get the helicopter ready and have a car ready in Philadelphia by the time we get there by helicopter."

The agents left to set things in motion. After they left, the President said, "I'm as excited as I have been in years."

"I promise you won't be disappointed sir," Eddie said.

They flew to Philadelphia in the copter and the car met them. Clarence directed the driver to the location and had the driver park the vehicle off the road, but far enough away from the site that they would not be able to see much, or hear anything.

Eddie and the President walked toward the grove of trees. "What do I have to do?" President Kelly asked.

"Not a thing. Tom will do it all."

They arrived and stood there. Suddenly they were in the presence of the alien. "I see you have brought another visitor," Tom said.

The President looked around, but mostly at the alien. "I see what you mean by the opaqueness," he said.

"Tom, this is our President, the man who rules our nation."

"I am pleased to meet you. What brings you back now Eddie?"

"Well, some things are happening that I can't account for, and I need you to shed some light on the happenings."

"Are you having trouble with the oscillator design?"

"Haven't even started it yet. We have gotten most of the computers back with the flaw. There are only thirty or so still out there, but something has happened that is very strange. We are now being watched at both of our locations by a large number of people, and I have no idea where they are from or who they are working for."

Tom handed Eddie the headpiece and said, "Let me see what you have?"

Eddie placed the headpiece over his head and after a moment, Tom told Eddie, "I think we have another player in the game. I believe someone else has stumbled onto a portal, probably someone of inferior intelligence, and my brethren is trying to figure out what is going on. He obviously found out about your computer and had someone check it out. I surmise that he asked the contact to check further and they found out about the oscillator project. I imagine he is now trying to neutralize both projects."

"Why would he do that?" Eddie asked.

"Because the computer will cause him to lose points, and he will guess that I or someone like me, is trying to neutralize another of our brethren."

"What do we do about it?" Eddie asked.

"You will have to deal with the physical threat in whatever manner you normally use. If you can find the location of the portal, you can take care of it the same way you are going to take care of Joe."

"In other words, you can't shed any light on this other than that it is probably the result of one of your kind interfering with our plans?"

"That's about it. You are going to have to deal with one problem at a time. The first thing you need to do is find out who the contact is, and that should not be all that difficult with your mental acuity. If you can either follow him, or get the location of the portal from him in some other manner, then you can deal with the second party later. I would advise that you watch the portal location as you are doing with Joe's to make sure there are no other players."

"Now, Mr. President, you have to decide if you want to use all your brain power or not. Tom can do that for you, and it will only take a few seconds," Eddie said.

"After seeing what you guys have gone through, I am not sure I want to."

Tom said, "Their problems will not befall you, only insofar as they need your help to deal with the current situation."

"Then let's do it."

Tom handed him the headpiece and he put it on his head.

"Okay, we are finished," Tom said.

"That fast. I didn't feel anything, and I don't feel any different."

"Believe me you will, and the change will amaze you," Eddie said.

"When are you going to start the oscillator?" Tom wanted to know.

"Soon, and I am going to have to come back again to get some specifics on the vacuum component. That's totally new to me. I am not sure I can even identify all the elements in it."

"Come back anytime. Remember, the specifications have to be followed precisely."

"Thanks for the visit," Eddie said as they returned to the grove of trees.

As they walked away, the President said, "I guess I am in your camp now, for better or worse."

"It isn't that bad. If you tried to use one of my computers now you wouldn't be able to visually keep up with your thought process. We had to modify one to slow it down enough to use. You are going to find your memory sharper than ever and you will recall anything you read forever. As a matter of fact, if you think about a book you read as a child right now, you will be able to quote what you read then. The hard part is going to be to keep people thinking that you are not very smart," Eddie told him.

"Thanks a lot," the President said with a laugh.

"I didn't mean it the way it sounded."

"I know what you meant, and I think you are right."

When they got back to the vehicle, Clarence said, "Well is everything resolved?"

Obliquely, the President said, "I think I understand the problem better, and you can count on me for total cooperation in the future."

They returned to the airport and took the helicopter back to Washington. It was almost three o'clock when they got back and the President said, "I need some more of your time after I try to get caught up on my schedule. Could I ask you gentlemen to come back around seven this evening?"

"Whatever you desire sir," Eddie said.

"I will send a car for you. One of these fellows will pick you up at seven. In the meantime, give some more thought to solving the personnel problems and finding the other person."

"I have already been thinking about that, and there might be a solution at hand. I will tell you about it this evening.

After they landed at the White House, they were taken

back to their hotel. In the room, Eddie said, "I want to capture one of the people watching each of the facilities and see if they will put us onto the man behind the activities. They appear to be druggies, or related to the trade, and if they use their own products, then we should be able to get them to talk. If not then there is sodium pentothal, or other substances to do the trick."

"How did the President react to Tom?"

"He was blown away, but he submitted to the headpiece. I think that was what he meant when he said we could count on his full cooperation.

"I would not have expected him to agree to that."

"Are you kidding? He has seen the results in us, and to pass up such an opportunity would take an awful lot of self control. He knows that we are not using our intelligence for harm, and probably figures that he can do as he wishes with his own now. I told him he is going to have to work at playing stupid, or words to that effect, and he got a good laugh out of that. There was apparently no outward sign of what went on?"

"None. It just appeared as if the two of you were holding a conversation, although now that I think about it, you didn't move much for a few minutes."

"I am going to talk to Major Dalton and to Bill. Is there anything in particular you want to tell them?" Eddie asked.

"You are better qualified to handle that than I am, so have at it."

Eddie called Dalton first. "What have you turned up?" he asked.

"I put the number at over forty, but they are low level. Probably pretty brutal, but not well trained. My people will be able to deal with them as they are now constituted. There was no evidence of weapons, although most of them probably had handguns."

"Do you think you can snatch one, preferably someone who looks intelligent enough to control the rest of them?"

"I don't think that will be a problem, but what do we do with him?"

"Smuggle him into the compound and keep him gagged and blindfolded. Also cuff him in such a way that he can't remove either the gag or blindfold. Clarence and I will be back there tomorrow and question him. Try to snatch him so that nobody knows he is missing."

"He will be waiting when you get here," Dalton said.

Eddie next called Bill. "What's the verdict?" Bill asked.

"Pretty much as we thought. Seems there's a third party involved. By the way, the President is one of us now, if you take my meaning?"

"No kidding! That surprises me in a way, but then it is only logical in another way of thinking."

"We are going back for a skull session with him later this evening. I have instructed Dalton to get one of the more intelligent looking of the watchers so we can question him when we get back. Have your troops do the same and smuggle him into the building, handcuffed, blindfolded and gagged. I want to compare the stories they tell."

"All the troops are here now, and the ones who went out last night say there are closer to fifty of them than thirty or forty."

"Do we have adequate manpower to handle them?"

"The Captain in charge of the troops say they can handle what's there now, especially with the weapons load delivered today. If the number grows he would feel better with more people."

"I will relay that to the President tonight. We will see you sometime tomorrow."

To Clarence he said, "Bill says the Special Forces guy says there are closer to fifty in Albuquerque than the earlier estimate of thirty. They are going to bag one and we can talk to him after we address the gentleman in Carson City."

"You know, this is the most excitement I have experienced since I was seventeen," Clarence said.

"And what happened that was so exciting when you were seventeen?"

"I lost my virginity. Things haven't been the same since," he said with a prolonged laugh.

"Let's get some supper. We don't have a lot of time," Eddie said.

They went downstairs to the restaurant and were shown to a table somewhat removed from the crowded area. They ordered and had a decent meal of salmon. At seven they were by the entrance waiting for their ride, which was very prompt.

They were taken to the Oval Office again and the President dismissed the Secret Service detail. That didn't mean that they were alone, because there was a peep hole in the door that allowed the protective detail to observe what went on in the room, but not hear the conversation.

"Okay, bring me up to date," the President said.

"We now have twenty two Special Forces troops in Albuquerque, along with a stock of heavy weapons pilfered from some federal armory that will be reported stolen in a few days when the inventory is done. All the troops are on official leave and are carried by our company as part time employees. There is an opposition force of at least fifty around the compound, but as far as we can tell, only armed with light weapons. The plan is to snatch one from each location tonight and have Clarence and myself question them tomorrow. I need to find the identity of the head guy to determine the location of the other portal. I believe it is only a single individual much as Clarence was for over forty years. This one is in touch more frequently I think, and is probably getting more help amassing the funds needed to do what he wants done. If we can get a general location it will be a big help."

"What else do I need to do from my end?" asked the President.

"Just make sure General Clarkson keeps providing what we ask for. I look at this as a genuine national security problem as what happens will have far reaching implications in the future."

"I am in full agreement with that, and I have authority to act unilaterally as I see fit to protect the nation. We will do what has to be done and sort out the repercussions later. Just for your information, I breezed through my appointments much quicker than usual. My grasp of concepts seems to be much sharper."

"That is one manifestation of the change. Another is that you are going to have a hard time curbing your desire to do things you know you can do, but nobody else can. It was that way with Bill and the AIDS cure. And the computer was an obsession with me right from the start. You might not have as hard a time as we did because of your advanced age, relatively speaking, of course," Eddie said with a laugh.

"I think I know what you mean. I look at things differently now, as if I really understand them for the first time. I have ridden in cars and flown in planes and never given it a second thought. Now I will be thinking about how the engines and electronic components work, and wondering why someone didn't think of what I am thinking to improve their performance."

"I see you have the essence already," Clarence said.

"I also will bet that you will not seek a second term as President, though you would be a runaway victor if you do."

"What do you mean by that?"

"It will be too restrictive. The office will prevent you from doing the things you really want to do and you will be miserable if you have to spend another four years here.

I don't know much about your financial situation, but you will never have to worry about money but rather how to curb your activities to keep from making money. I know that sounds strange, but believe me, you will recognize it before you finish this term."

"If that's the case, it will give me time to groom someone for the job."

"That's true, and your reasoning will be a lot sounder now in the things you look for in a successor."

"Make sure everyone in the circle has my cell phone number," he said.

"They already do. We will be going for now. I will keep you posted as events develop. You should know that the location we are going to use the EMP gizmo on is in Yosemite National Park. Since we don't know what the actual effects will be, when we are ready to do it, the FAA should clear all air traffic over the area. That's going to be a few months away, but is something that we know will be required."

"You think I will understand the concept of this system better now?"

"Yes sir. You will be able to look at the blueprints and see exactly how it will work, though the intricacies will elude you, just as they do us. That's why I am going to have to go back to Tom as we progress, to make sure we are doing it right."

"Well, good luck, and keep in touch regularly," the President said as he stood up to dismiss them.

Chapter 27

When Eddie and Clarence got back to Carson City, Major Dalton led them to a corner of the hangar where a seedy looking Latino was bound and gagged. His pants were soiled with urine and he smelled very ripe. Eddie mimed retching as they neared the man.

The gag was removed and he gulped a good lungful of air. He then spouted a stream of Spanish at them. Eddie waited a few seconds and said in English. "You are in America and we speak English here."

The captive again exploded with a stream of Spanish expletives.

"Put the gag back on him. He will be more cooperative after he has another twelve hours to think it over."

"No, No, what you want with me?"

"I just want to know why you are watching my business facility, along with many others. You don't even know what we make, and if you were thinking of shaking us down, we have more security people than you have hoodlums."

"I was not watching your place. I was looking for a good place to cook meth."

Eddie humored him. "For yourself or for sale?"

"Some of both. I am taking over the territory soon and need to know the area."

"Who do you work for?"

"I don't work for anyone but myself."

"I am not stupid. I know how the cartels work, and there is no such thing as an individual operation. They close those down very quickly. Now who do you work for?"

"I work for a man named Carlos Estrada. He controls many areas for distribution in the western states."

"I have never heard of him. He must be way down in the pecking order."

"I think he just took over recently, and he has much money, and handles all the transactions in the western area except California."

"What's your name?" Eddie asked.

"Ramon."

"And where are you from Ramon? I know you are not from around here?"

"I recently moved here from Texas?"

"What part of Texas?"

"Odessa."

"What enticed you to make the move to a small city such as this one?"

"The potential for profit is much better with the gambling and Lake Tahoe close by."

"How many men did you bring with you?"

"Only four of us."

"And you know nothing about the other forty or so that are all around this facility?"

"I don't know nothing. I am just following orders."

"How long have you known Carlos Estrada?"

"Off and on for a few years. We used to do business in a small way."

"Does that mean that he was your supplier?"

"Why are you asking all the questions? You are not the police. You have no right to hold me."

"Because you have blundered into a top secret government project, and the government takes an even harder line with that than with your drug dealing. I suspect that you could be kept in isolation for several years without anyone knowing. No lawyers, no trials, just the fact that you were trying to breach this facility will be enough to keep you under wraps for a long time. On the other hand, we may just decide to get you permanently out of the way so we won't have to worry about feeding and housing you, not to mention what the guards would cost. What do you think Major, should we just have a couple of the men take him away and make him disappear?"

"I think that's the easiest solution. We can pick them up one at a time and before you know it the problem will be gone. We still have the back hoe, so we can dig a hole large and deep enough to hold twenty or thirty right here on the property. With the privacy fence no one will be the wiser," Major Dalton said, catching onto Eddie's ploy.

"Go ahead and set that up. I don't think this one is smart enough to know anything anyway," Eddie said and turned as if to walk away.

"Wait a minute. You can't do that," Ramon practically screamed.

"Why not. Look around you; do you see anyone to stop us?"

"What do you want to know? I don't know much, but I will tell you all I know."

"Go back to the number of people with you," Eddie said.

"There are forty six altogether. Most are from Texas and New Mexico. We picked a few up in Vegas. Carlos said to watch this place until he told us otherwise, and that he would provide us weapons before we did anything else."

"Tell me about Carlos. Was he a supplier for the cartel when you knew him earlier?"

"Yes, for Escobebo. I had not been in close touch with him for many months, and I think he had the habit bad, but he somehow came into a lot of money. I don't know if he ripped off one of the larger stashes of the cartel or what, but he seems to have a lot of money. When he contacted me and offered this job I checked to see if he was on the cartel's untouchable list. He is not wanted by them, so he must have gotten the money somewhere else. He came to me and offered a large sum of money if I would put together a group and come here. I assume that his plan is to take the place away from you. I don't know if he thinks it is an independent drug operation or what, but for the kind of money he was offering I couldn't turn him down."

"And now that you see what you are up against, what is your judgment?"

"I think we are in over our heads and Carlos doesn't know what he is into."

"Where does Carlos normally hang out?"

"Mostly in the south of Texas. He has lived in El Paso, Odessa and Houston that I know about, but probably other cities as well."

"How do you contact him, or does he initiate the contact?"

"I call a cell phone number when I have something to report. He can call me as well, although he has not done so in the past four days."

"Do you have his cell phone number programmed into your phone's directory?"

"Yes, as number one."

"Who is in charge of the others watching this building?"

"I am in overall charge, but I have sub lieutenants for each ten men."

"What's their reaction going to be when they find you have disappeared?"

Seeing a glimmer of hope, Ramon said, "They will only think I went to get a bit of recreation if I am not gone too long."

"What do you think Major, shall we round the rest up and get rid of them all or try to let Ramon here take them away?"

"I will have to give it some thought. Let me talk to my men and I will let you know in a couple of hours."

"Put the gag back on him?"

After they left the area where Ramon was stashed, Eddie asked, "Where are his possessions?"

Dalton took Eddie to the office they had set up and opened a desk drawer. A wallet and cell phone were the major contents.

"Did you find out the number for this phone?"

"No, but all you need to do is call your number with it and check the caller ID. What do you have in mind?"

"I'm wondering if we provide the numbers of this phone and to Carlos's the NSA can use the GPS feature to get a location for Carlos?"

"I'm not intimately familiar with their capabilities, but it stands to reason that they can."

"I'm going to take the phone and see if I can work something out. If you don't want to mess with Ramon, smuggle him outside and turn him over to someone for safekeeping. I am going to talk to another one in Albuquerque and see how close his story is to Ramon's, then decide where we go from there."

Clarence had been silent throughout the charade. Now he asked, "Do you want me to stay here, or go with you back to Albuquerque?"

"If you don't mind, I want you to stay here. We need someone in the loop at both locations pretty much full time now, and the computer programming still has to be done down there. I will call you when I figure this out."

On the way to the airport Eddie called the President's cell number. "What do you need Eddie?"

"I need to have a phone located by the GPS feature embedded in it. Can the NSA do that?"

"I'm sure they can, but I will check and get back to you in a few minutes. You have a phone number for the man we are looking for?"

"Yes, and a general location. He is probably in south Texas. If it looks like this has a chance I plan to take a trip down that way, and another small force might be needed."

"I'll call you back in a few minutes."

Eddie was on the plane getting ready to taxi when the call came. "We can get the location if they have the phone number and can intercept the call. Knowing the general location will help them narrow down the search area. I am going to give you a phone number. Call and give them the particulars, then let them know approximately when you are going to initiate the call. Once they isolate the number, they can give you the location instantly."

"I want to be in a position to react when we do this to try to get a location of the portal."

"Call if you need anything else. General Clarkson can give you more people if needed. I have talked with him again and stressed the importance of what you are doing."

Eddie called the number. The person who answered the phone did not give a name but quoted the phone number. "My name is Eddie Casteen. I was told to call this number to set up a procedure for locating a specific cell phone."

"Yes sir. I am aware of the tasking. If you can give me the number which interests you, we will put it on the watch list. If it becomes active we will get the location, but I got the impression that you might want to do this at a specific time."

"I know that the phone is somewhere in south Texas, but not any closer than that. I would assume it is in a pretty desolate area based on other known factors. What I would like for now is that you keep track of his location when he uses the phone until I can get the other parts of the plan in motion."

"Are you interested in what the party discusses?"

"It might give me some more insight, but is not necessary to what I have in mind. Just do it like you usually do, and please keep it as closely held as you can."

"Let me have the number and we will get it set up. Can I reach you at the number you are calling from at any time?"

"Yes, you can, and I thank you for your cooperation."

"Everyone cooperates when the call comes from where this one did."

Eddie now had to decide how to deal with the situation if he located Carlos. He would have to have him followed in order to determine where the portal was located, and he knew he couldn't do this on his own. Since they didn't expect any attacks on the facilities right away, he thought it might be prudent to take part of the Special Forces troops from the Albuquerque facility and head for Texas.

When he arrived back at his office, he found another prisoner of the same ilk as the one in Carson City, but a bit cleaner.

Before he talked to him he got his group together with the leader of the military troops. "I want all you guys to hear this, then I will tell you my proposal for dealing with the situation. Feel free to jump in at any time if you have questions."

He reiterated what had happened at Carson City, and the responses he had gotten from Ramon. "I believe he was truthful, and that he thought we would do exactly as I told him. He saw the number of troops and the armament and knew that his group wouldn't have a chance against them. I think he was genuinely afraid that we would wipe out his entire group and cover it up."

"I called the boss, if you take my meaning, and asked to have NSA get a location on the cell phone that Carlos is using. I have already been in touch with them and set it up. Before I call General Clarkson, I wanted to make sure we had enough troops here to protect the place if I take about a half dozen of these guys to Texas. That's the broad strokes. Any comments?"

"I assume you want to keep the production line working here?" Beth asked.

"If we can do so without endangering the people, then yes, but not at their peril."

The Captain who was sitting in on the meeting said, "I know there's something here I am not aware of, but why is it so important to locate this Carlos?"

"Number one, he is the guy bankrolling all these hoods we are faced with, and number two, we absolutely have to know where he is hanging out so we can deal with him and his benefactor, or the problem will just keep cropping up with new players in the field."

"I wouldn't feel comfortable taking six men away from here without replacing them. The force outside is large enough to cause a lot of problems, and some just might manage to get past us with a concerted effort and a reduced defense."

"I don't believe they will act within the next forty eight hours, and we can get additional people here by then. But, before we do anything, I want to talk to the captive and see if he verifies what I got from Ramon."

He and Bill went to the office where the prisoner was being held. He was trussed up the same way Ramon had been, and the gag was removed, but the blindfold left in place. Eddie started the questions pretty much as he had with Ramon, and the results were pretty much the same. Defiance, and belligerence. Eddie took the blindfold off the man and asked his name and where he was from. He gave an obviously fake name, so Eddie pulled out Ramon's cell phone and scrolled through the names until he came upon one that seemed likely and pressed the call button. The captive's phone was on the desk and started to chirp.

"Well," Eddie said, "That tells me your name is Jesus. You sure aren't much like your namesake are you?"

"Whatcha talking about man?"

"You know, Jesus of Nazareth, or haven't you ever read the Bible?"

"I go to confession regular like, and I even go to Mass sometimes," he responded.

"Would you tell me who you are working for and what you are being paid?"

"I ain't working for nobody. I do my own thing."

Eddie had an idea. He excused himself and motioned for the two girls and Bill to come outside in the hallway with him. When they were outside he said, "I don't know why I didn't think of this before, but do you realize that the computer can be used as a tool for interrogation?"

Bill caught right on, "Whatever he thinks of will flash on the monitor. We just have to make his thoughts go where we want them. I like it."

The girls understood as well and Beth went to get a computer.

"Captain, you are about to witness the very latest in interrogation techniques. It might be something that will come in useful to you in the future," Eddie told him.

Beth came rolling a cart with the monitor and the headpiece on the shelf underneath. They sat the prisoner up in the chair at the proper angle and bound his hands in position with the handcuffs and held his head in place with a shoelace tied across his face and secured to the headrest of the chair. They left the gag in place since they would not need him to verbalize.

When everything was ready the headpiece was placed on his head and the computer turned on. The monitor was turned away so that he could not see what was happening on it. It took the usual few seconds to select the frequencies and came to life.

"Now," Eddie said, "I am going to ask you the questions again. You don't have to answer, so the gag will stay in place."

Everyone was in a position where they could see the monitor, except the man hooked up to it. As Eddie asked the questions visions appeared on the screen.

"How many men do you have here Jesus?" Though he didn't answer an image of a large number of men appeared on the monitor. "I need you to think of the number Jesus," Eddie said.

The number appearing on the screen was eighty.

The Captain had not seen one of the computers in action and was totally amazed.

"Now who do you work for?" Eddie asked.

An image of a Latino appeared on the screen. The image was good enough that those who glimpsed it would be able to visually identify the man. "And what is the name of the man? Spell it for me."

Carlos appeared on the screen.

"Where did you meet Carlos?"

A picture of a smoky bar appeared on the screen.

"I need to know the city," Eddie said, and the image changed to a landscape. They could not identify the city from the view, and Eddie told him to spell it.

He spelled out Odessa.

"I assume this is in Texas and not Russia," Eddie said.

"How much did he pay you for your part in this?"

Again the dollar figure flashed on the screen. This idiot was not even smart enough to try to think about something else. When Eddie asked the questions he just thought the answer.

Eddie even got the location of the bar where they had met.

"Anybody need to know anything I missed?" he asked.

"Jesus, how long are you supposed to watch before you take action?" Beth asked.

The screen remained blank. "I don't think he knows," Eddie said.

"Were you waiting for a signal to attack the place?" The screen still remained blank."

Eddie suddenly had a premonition. "Beth, is this one of the recall units?"

Beth caught on right away. "It wasn't supposed to be, but let me check." She removed the headset from

Jesus head and checked for the marking on the new ones and didn't find it.

"I think this was the original prototype. Jesus, is this what was supposed to happen to the entire population?"

"I'm afraid so. I don't know if it will do any good, but I think we should have Jesus get a brain scan, but I think you will find that he is in a vegetative state. Take the shackles and ropes off."

Jesus was conscious but did not respond to any sort of stimuli. "I guess Joe got tired of waiting. I need to pass this to Clarence and the other one as well."

The Captain was totally befuddled. "What just happened? It's like his mind just winked out?"

"That's exactly what happened. Can you smuggle this guy to a hospital Bill, and set up a scan to see if there's any way to reverse it."

"I think it is going to be beyond our knowledge level, but I will see what can be done. You are thinking of the twenty something still out there?"

"Yes, but that is better than the millions planned."

Eddie called Clarence right away. "Hold onto your hat pardner; Joe just activated his plan for the first computers. We were using one to question one of the local hoods and he just winked out in the middle of the session. He is alive, but non-responsive to any stimuli."

"I guess that proves that Tom is the good guy."

"It would seem so. I am going to beef up the security here. It seems there are eighty of the hoods here, and more will probably arrive before they decide to do anything."

Eddie next called General Clarkson. "General, I am going to take about half a dozen of the troops from Albuquerque with me on another part of this mission. The bad guy's number eighty now. Can you send another twenty or so to beef up the security here? The Bad guys don't have any heavy weapons yet, so the numbers are the primary concern."

"The President told me to give you anything you asked for without question. Do you want them in Albuquerque?"

"Yes sir. I am going to take the Captain and five more with me to Texas, somewhere near Odessa. If I find what I am looking for, I will need to set up a perimeter around a location yet to be identified until we can complete the project we are working on. I will let you know more about that part of it later."

"I'll have the additional troops on the way within the hour."

Eddie next called the President. "I seem to be very popular with you today. What now?" asked the President.

"The location I am looking for is near Odessa, Texas. We were using one of the computers as an interrogation tool, which by the way works great, and the guy we were questioning suddenly went blank. He is non-responsive to stimuli, but is otherwise normal. I am having Bill run a brain scan on him at a local hospital. It seems Joe activated his plan, and I am thankful that we got most of them off the market early enough."

"At least it verifies Tom's take on the situation."

"That's exactly what Clarence said. I asked General Clarkson for some more troops. If I find the portal we are going to have to set up a perimeter around it, and I think troops are the best method to do It.?

"I agree. How long do you think it will take to build the oscillator?"

"I just don't know. We have most of the raw materials to start, but taking care of the opposition is the higher priority right now."

"I also agree with that. Call me when you get to Texas. I will give the matter some thought and might be able to help out with the problem."

Eddie told the Captain, whose name was Leyland, to round up five of the troops and pack a bag for a few days.

He then called the pilot and told him to file a flight plan for Odessa, Texas.

They were at the airport a little more than an hour later and boarded the plane for the flight. After they were airborne, Eddie got together with the troops and explained as much as he could. "Captain Leyland and I have seen what the man looks like that we are trying to find. We know that he frequents the Odessa area, though that may not be his home turf. I have his cell phone number, and as soon as we are set up down there, NSA is going to try to use the GPS function in his cell phone to get us a location. Keep your weapons with you at all times. I don't know anything at all about this guy, except that he has deep pockets and has hired a lot of unsavory people. I have to assume that he has some personal protection, and we will have to evaluate that after we get eyeballs on him. I don't want to kill him, but follow until he leads us to the location I am searching for."

"How do you want us to deploy?" the Captain asked.

"I think we should rent at least four cars. You guys can team up two to a car. We also need cell phones to communicate. Make sure everyone has the numbers plugged in for everyone else. Do we have a Spanish speaker in the group?"

Two of the men raised their hands. "What I would like to do is have one of you call Carlos after we are set up. You need to think up some excuse for having Ramon's phone, and remember he is in Carson City."

"I will pick up binoculars when we arrive. I want one set in each car. You might also think about stocking up on some snack food and water for each car as well."

"Now what have I missed Captain?"

"Rules of engagement. When are we allowed to use our weapons?"

"Self defense only. I don't anticipate having any gunplay until later in the proceedings, but if you are attacked, you can certainly defend yourselves."

There were no other questions and the plane landed in Odessa on schedule.

Chapter 28

After they rented the cars and did the shopping, Eddie led the group to the location of the bar Jesus had identified. It was unlikely that Carlos would be there at that time of day, but Eddie wanted to check anyway. He might be able to question the bartender or some patron about Carlos' habits.

That ploy didn't reveal anything helpful, so Eddie led the group out of town to a desolate area and called the NSA contact. "I am near Odessa, Texas, and we are about ready to initiate the call," he told the man.

"Give me five minutes to get to another area. You don't need to call back, just give me the time then make the call."

Eddie waited for seven minutes and had the call placed.

The man making the call asked, "Is this Carlos?"

He answered in the affirmative.

"This is Hector. I have Ramon's phone because he had to go to the hospital. He was having some bad problems with his vision and I took him. They think he might have had a stroke or something like that, and he told me to call you and ask if it was okay if I took over his position. He told me to tell you that nothing unusual has happened here but wanted to know if we were going to get heavier weapons before the assault."

Carlos rattled off a stream of Spanish, and it was fortunate that Eddie had the foresight to have a Spanish speaker make the call.

The man replied in Spanish, and the conversation continued in Spanish. Eddie didn't know what was being said, but got a thumb up from his man.

After the conversation, the man recapped for all of their benefits. "He wanted to know how I knew Ramon, and if Ramon was going to be able to participate. I told him I was not sure until I went back to talk to Ramon. I didn't give him the name of any hospital, but it wouldn't surprise me if he checks the hospitals in the area."

Eddie called Major Dalton. "Ramon is in the hospital somewhere with symptoms of a light stroke. Can you make sure that information can be verified at one of the local hospitals?"

"I don't know. I suppose if nothing else, I can have a man impersonate him and check into the hospital. I will figure it out. I assume you want this done yesterday?"

"There's an outside chance that Carlos might check the story we told him, and I want someone to verify it if he does."

"I'm on it. I assume we are not expecting any fireworks real soon?"

"It doesn't look that way, but remain vigilant," he said with a chuckle.

"Always. Talk to you later. I will brief Clarence, so you won't have to call him."

A couple of moments later, the NSA man called back. Got a location for you. It is not too far from Odessa. Looks like you made a right choice." He quoted him the geographical coordinates. "We have his voice in the recognition program now, and anytime he makes a call and the computer recognizes his voice, the call will automatically be recorded. I have modified the program to get his GPS location every time he uses his phone."

"Thanks. We are now going to run down the location you gave me and see if it yields anything."

"Call back if you need anything else."

Eddie had brought a hand held GPS, and now programmed the coordinates into his unit. It appeared they were on the right side of town and Eddie led the caravan toward the location. About twenty miles south of Odessa, a farm road intersected with the highway. No houses could be seen, and it appeared the location they were seeking was a good ten miles off the road. Eddie took the turn off and drove a couple of miles down the dirt road. As he glanced in his rearview mirror, he noticed the dust cloud following the vehicles. It would not do to get within sight of the location they were seeking with the evidence of their travels behind them. He stopped and everyone got out of the cars for a confab.

Eddie told them, "This is not going to work. Did you guys notice the dust cloud we are making?"

"Yes sir, but we figured you knew what you were doing."

"According to the position the NSA gave me, we are within five miles of the location now. I suggest we find some place to stash the cars within the next couple of miles and proceed on foot."

The four cars moved ahead at no more than ten miles per hour to keep the dust trail from giving away their presence. The land was very flat, but there were washes and lower areas in places. Eddie chose one that looked like it made a turn a short ways off the road, and slowly made his way around the small hillock. It was not much, but got the cars out of sight of the road. Once they were all parked one of the men went back and obliterated the tracks they had made with a jacket.

Eddie told the group, "You guys are the experts at the next phase, so you take the lead Captain and keep an eye out for any surveillance he might have out."

Eddie passed the GPS device to him so he could keep them going in the right direction. The Captain said, "Let's go back to the road and stay near it until we get closer. There's no sense dodging cacti this early in the game."

The group walked for the better part of an hour without seeing any living thing. Eddie thought it odd that they didn't even see any cattle.

As they got closer to the GPS location given them, there was still no sign of anything, especially human habitation. The closer they got to the location, the more careful they became. The Captain had the men spread out and continued the approach. They were within a half mile of the location and still had no indication of life of any kind.

Eddie said, "Let me see the GPS Captain?"

He handed it to Eddie, who satisfied himself that they were at the right location. "Let's you and I go forward a little way and see if we can get a clear sight line to the location."

"There's a bit of a hill over that way," the Captain said, pointing to his right.

"Then let's head that way. Have everyone else stay put until we get back.

The two made their way to the raised area. The location provided a view of the area in question, but there was nothing there to be seen. Eddie's first thought was that the coordinates had to be wrong, but he had a second thought. What if Carlos had just been in touch with the alien when they had made the call? If that was the case, then they had to be very close to the portal.

He scanned the area with the binoculars he had around his neck. He could make out tire tracks where the roadbed ran, and an area that looked as if it had recently had traffic. If this was the location of the portal, then all they had to do was wait for Carlos to come back to make contact, and they would have the location fixed. The only problem was that Carlos might not make contact for a very long time. They couldn't afford to go much closer for fear of being detected by the alien.

To the Captain, Eddie said, "I think we may have gotten lucky."

"How so. The guy is not here?"

"It's not so much Carlos that I am looking for, as the place where he communicates, and I think this is it."

"You're not making a lot of sense," the Captain said.

"I know, and I am sorry for that, but this is a very unusual circumstance. Only six people know about what we are doing, and the President is one of them. I am going to call him right now, and if you are willing, I want to ask him to put you in charge of the people who are going to be watching this site for as long as three months. If he agrees I will tell you what it is all about. You will need to know the importance and strict guidelines about the assignment, and I don't think I can impress you with the urgency of the task without telling you more about the situation."

Eddie took out his phone and called the President. "Did you have any luck?" he asked by way of salutation.

"I think we might have. He answered the phone from a remote location outside Odessa, Texas. We are at the location now and there is nobody here, but there are signs of recent habitation, at least evidence that a car recently traveled here. I have Captain Leyland and five other troops with me. We are going to have to get Carlos to come back here in order to pinpoint the exact location, and I don't know how long he will wait between contacts. I need someone here who understands the situation and can control all movement. I propose to brief Captain Leyland enough so he knows the stakes, and how to protect this area, because I need to get back to Albuquerque."

"Let me talk to him for a moment."

Eddie handed the phone to the Captain. "The President wants a word with you."

Captain Leyland couldn't believe that Eddie was talking to the President on a cell phone, but took the phone being handed to him.

"This is Captain Leyland," he said.

"This is your Commander in Chief and I am about to issue you a direct order. I want you to listen to what Eddie says, and do exactly as he tells you to do. You are not to reveal what he tells you to anyone, I repeat, anyone else. Is that clear?"

"Yes sir."

"Put Eddie back on."

"Tell him what you have to, but try not to get into specifics about cause and effect, if you know what I mean."

"Yes sir. I am going to call General Clarkson again and ask for more troops after I talk with Captain Leyland and get his input."

"Keep me posted."

Eddie hung up the phone. "What I am trying to locate is a portal that extraterrestrials use to communicate with humans. I believe we have found one, because Carlos is the contact for this one, and he was here when he took the phone call earlier."

"You mean like beings from another planet?"

"Yes. I know it sounds too weird to believe, but I have been in contact with them and I know how it works. They do not have the ability to harm the earth on their own, but they can influence humans to do the dirty work. What you saw in Albuquerque with the interrogation is an example of what they can do through humans. I designed and built the computer that an alien showed me how to build, except he put a flaw in the design that allows him to do exactly what he did to Jesus. I found out about the flaw when we had almost ten thousand of the things on the market, and had to redesign the system and recall all the units. We got all of them back but about twenty. What that means is that twenty people who use those computers are going to end up in a vegetative state, just like Jesus. That is better than the thousands or even millions that could have been affected if we had not found out about it in time. Now, the facility in Carson City came into existence to build a weapons system, for lack of a more descriptive term, to destroy the portals through which the aliens communicate. The only drawback is that the portals are not visible to us and the weapon has to be precisely directed in order to be effective. When Carlos is in contact there is no outward sign, other than there will be very little movement on his part during the contact."

"And you think the portal is down there," he said, pointing down the hillside.

"Exactly. Now the alien can sense human presence within fifty to one hundred yards from the center point of the portal, so we can't approach the location below. What we have to do is get enough troops to seal off the area completely, without anyone knowing they are there, and wait for Carlos to return and lead us to the precise location."

"I can't stay here to help with this. That's why I asked the President for permission to tell you. Whoever is in charge here has to know what we are protecting, and the consequences if we fail. The design for the system to take care of these portals, and yes I used the plural, was given to us by another alien, and we have been stockpiling the raw materials to build it in Carson City. Between the computer and the other unit Carlos has apparently given this alien enough information to cause him to want to stop both projects. That is why all the druggies and low life are showing up at both locations."

"You are not going to believe this, but I don't doubt your word for a minute. I have believed for a long time that there were ET's, and that they had visited our planet in the past. I read everything I can get my hands on related to the subject, and the opportunity to participate in something like this is a lifelong dream."

"Remember the need for secrecy. If word of this got out it could easily cause a panic, and possibly result in a tremendous loss of life."

"I understand, and I will do what has to be done."

"How many men and what weapons will you need to set up a perimeter in all directions a quarter of a mile from the location we are looking at down there?"

"We have to remain totally covert for this?"

"Yes, and nobody comes or goes except Carlos. You and I are the only ones who know what he looks like, so you personally, will need to be in a position to identify him if and when he shows up."

"Do you have any indication as to how long before he will show up?"

"None whatsoever. It could be days, or weeks, so plan accordingly."

"I would set up observation posts about a hundred yards apart all the way around the perimeter you talked about, two men per post. Each post will be camouflaged with netting and native vegetation. They should be delivered by helicopter about a mile in an easterly direction. I will meet them and lead them in from there. In the meantime, I can have the troops here now scout out the best locations and mark them for the coming troops. Do you know where the troops will be coming from?"

"No. Do you have some preferences?"

"I would like to have Special Forces. They are better trained for this type of activity, and they know how to keep their mouths shut. We can fly them into any airport you want to and copter them in from there. If you can arrange for a military chopper or two from Fort Bliss, or even Sheppard AFB it will work out better."

"What's the total number, and how do you want them armed?"

"We are not expecting anything heavy here are we?"

"No, probably their normal armament will be good enough for anything you will have to face here. I don't really expect anyone to show up other than Carlos."

"I think fifty men will be required to make sure we have eyeballs on the total perimeter."

"Will Special Forces have that many available?"

"I would think so. They always have a ready reaction force of at least that many."

"Then let me call General Clarkson and set it up."

Eddie made the call. When the General came on the phone he asked, "What can I do for you this time Eddie?"

"I am near Odessa, Texas and we have an area that needs to be placed under constant surveillance. Captain Leyland, who is with me, says he needs about fifty men to do the job. I don't expect any rough stuff here, but it is essential that the surveillance force not be detected. The Captain says Special Forces would be best, and nobody over the rank of Captain. Leyland is going to be in charge so I don't want any cause for dissension in the ranks."

"I have dispatched more people to Albuquerque already. How do we get the people to you?"

"The Captain suggested flying them into either Fort Bliss or Sheppard AFB, then helicopters to drop them off. What I suggest is that you call Bragg and tell them Leyland will be calling with the particulars. He is going to need radios and other stuff, which he can better tell them than you or I."

"Then I will do that. Is this going to continue to grow?"

"It depends on the opposition, but I believe we have enough manpower to handle all the situations now."

"I would say nice to hear from you, but I don't want to lie," the General said as he broke the connection.

"Give him time to make the call, then call Bragg. The skids should be greased by then."

"What about the cars?"

"Drive them as close as you can to here and hide them. You will eventually have to go back to Odessa."

Captain Leyland rounded up a couple of his men and with Eddie, walked back down the road toward the cars. Eddie drove back into Odessa and to the airport. He was back in Albuquerque later that evening. He asked Bill about the scan on Jesus.

"Everything is blank. He has absolutely zero brain function."

Eddie then told them about the trip to Odessa. "I think we got lucky with the phone call and caught him near the portal. I have security set up around it. I had to tell the Captain more about it to make sure he doesn't muck it up. We are getting fifty more troops to surround that location. Leyland knows what Carlos looks like, and what to look for to pinpoint the portal location."

"I called Clarence on the way here and briefed him. I think I need to go to Carson City and try to get started on the oscillator. If necessary we can put off the computer deliveries until things settle out with the other things."

"Are we all going to Carson City to start the work?" Beth asked.

"I think we have to leave someone here, and I would suggest Bill," Eddie said.

"That will work. With the additional troops coming, I can continue to program the computers at a slower pace. If something should happen, I will be available."

Eddie was going to fly to Reno that night, but decided to wait for the arrival of the additional troops before he left.

He just had too many balls in the air right now, as the juggler said. And there was nothing he could do until Carlos initiated the action, or showed back at the portal site.

He flew to Reno that evening and got together with Clarence at the hotel. He brought him up to date on the day's happenings and told him that he was going to start building the oscillator right away.

"We also need to make sure we have enough troops on hand at all times. I will get together with Major Dalton and set it up to have at least half on hand at all times," Eddie said.

Eddie also called the people watching the site where Joe was. There had not been any activity there so Eddie assumed Joe just decided to activate the problem with the earlier computers on the outside chance that it would be effective.

The housing units had been delivered during the week on semi trucks. Dalton had placed the ones that were to remain outside strategically to provide cover for his troops if they got into a gun battle with the opposition.

Dalton had sent people out to spy on the druggies, and had learned that they were wondering what happened to Ramon. Someone else was running the show now, and from the accounts of infiltrators, didn't seem much more competent than Ramon.

Many of the force was still using drugs on the job. It didn't appear that they were much of a threat, but when you had more than fifty people with guns, you could not be sure of controlling them.

Major Dalton had devised a plan to have part of his on duty force flanking the druggies outside the compound. If they should attack, then they could catch them in a crossfire and handle them more easily.

There was still no sign of heavier weapons and they were thankful for that. Also, the compound was in an industrial area and there would not be a lot of civilians in the line of fire if anything happened.

All in all it was about all they could do until the other side made a move.

Chapter 29

Eddie was at the compound early the next day. He and Clarence started to look over the design to figure out where to start. It appeared that the armature portion would be the bulkiest and most tedious part of the work. They enlisted several of the on duty security troops to do the heavy lifting, and several of them volunteered to help with the wiring. By the middle of the day, Eddie and Clarence had a good feel for the process and they worked on two of the armatures at once.

They ran into problems with the weight of the wiring and had to devise a way to hold them in place long enough to do the work. Eddie sent one of the men out to purchase four chain falls from automobile repair shops to stabilize the part they were working on.

It appeared that they would be able to build a complete armature in a couple of days, so Eddie started thinking about how they would hold it in place inside the housing. He mentally reviewed the plan Tom had given him and could only see two places where it could possibly be attached, and those didn't look strong enough to support the weight of the metal in the design.

Clarence and Eddie talked it over and looked the design over again. Clarence finally said, "Could the magnetic force from the center component exert enough force to hold the armature in place?"

"I don't know, but it is a thought worth considering. I don't think the alternatives are workable. The metal is simply not strong enough to hold the weight. If the core exerts negative magnetic force, it could keep it in position."

"Let's go with the next component and leave that problem until later. If we have to go back to see Tom, we might as well try to find out what else is going to give us problems."

Beth and Cindy were scheduled to arrive the next day, and two more minds might make the problem solving process a little easier.

Major Dalton came to see Eddie late in the evening. "I have some bad news. My spies have detected a shipment of automatic weapons coming to the opposition. I think they might be getting ready for the assault."

"Alert all the troops. I will call Albuquerque and let them know. I expect they will hit both places simultaneously."

Eddie also called the President. "The bad guys just got a shipment of automatic weapons. We expect them to act anytime from now forward."

"Try to keep the collateral damage to a minimum. Since you are in an industrial area that might be easier than would normally be the case."

"Yes sir, and we are prepared. I am going to alert our people in Albuquerque as well. I think they will try to act at the same time."

"Keep me posted."

Major Dalton called the hotel where his people were staying and told them to report singly or in pairs, and for those assigned outside the compound to get into position, with their rifles.

By ten o'clock the entire force was either in the compound or at their positions outside. Dalton had men manning the scissor lifts with automatic weapons, and those would have the best sight line as the force approached the compound.

"Wait for my signal to open fire," Dalton told them. "And be careful of your line of fire. We don't want a lot of dead civilians when this is over."

The phone call to Bill in Albuquerque had been unnecessary. The troops had detected the same thing there and were preparing for the worst. There was no way the confrontation would not end up a slaughter of the ill prepared storm troops. They were simply no match for trained soldiers, no matter how they were armed. Add to that the fact that probably half of them were on drugs and the equation was skewed even more.

They waited patiently until after midnight. A cell phone rang, and the caller told them that the group was getting ready for the assault. Carlos must have offered a great deal of money to get as many men as he had arrayed against them. Most in that culture considered themselves pretty bad, but the opposition was usually unarmed and either kids or older people. They were about to find out that they weren't so bad after all.

By now Carlos had to suspect something with Ramon being out of touch for so long, and he had sent another man he trusted to lead the group.

Just before one AM they detected movement on all fronts. They waited for the assailants to make the first move. Major Dalton was hoping one of them would get an itchy trigger finger and give his people the provocation they needed to take care of them.

He wasn't disappointed. As soon as the group in front of the blockhouse got within fifty yards one of them opened up with a sub-machine gun. The projectiles did not penetrate the bullet proof glass, and the trooper lying on the roof took care of the shooter with a single round.

Other rapid fire could be heard all around the building now, though the assailants could not see what they were shooting at with the privacy lacing in the chain length fence. It also appeared that the idiots had not thought of a way to breech the fence once they got to it. The shooters on the scissor lifts at the building corners had the best line of fire, both from the safety standpoint, and the visibility.

The lights had been on in the compound but now were turned off. The troops had night vision goggles and donned those after the lights were extinguished. The assailants were now at a further disadvantage.

The troops inside the compound moved to the fence and removed enough of the slats to get a line of fire and within minutes a cacophony of gunfire could be heard.

The battle was as one sided as it should have been, and very few of the assailants escaped. The ones who tried were cut down by the troops stationed outside the compound.

No one tried to surrender. There was not even enough time for them to realize what a terrible mistake they had made. Bodies littered the area outside the compound and soon the gunfire died down to a trickle.

The shooters in the high perches took care of the remnants and inside of twenty minutes it was over. The compound lights were turned back on and the troopers went out to assess the damage and care for the wounded.

Not many were left alive. There were over fifty dead, and only about half a dozen that looked as if they might survive with medical care.

Soon sirens could be heard as police started to arrive on the scene. Major Dalton had alerted them to the possibility of an attack and asked them to delay their response when it started so they wouldn't get caught up in the action.

The first car to arrive contained the Chief of Police for the city, and Major Dalton greeted him as he got out of his police cruiser.

"I think everything is under control. There are several wounded that need hospitalization. Looks like the body count is over fifty."

"What are you guys doing in there that someone would put this kind of force against you?"

"I don't even know. I was told that the project was top secret and that there was a threat. My unit was sent to protect the place, and that's what we did. I don't know how to spin this for the press, but with that many bodies, they are going to have to be given something plausible."

"It looks to me like rival drug gangs had a falling out. Some of these people are known to be associated with the trade, and this is the gateway to Lake Tahoe, so that would be plausible."

"The warehouse next door is vacant, so that could have been where the battle took place," Major Dalton offered helpfully.

"Exactly. That is obviously what happened. Have your men gather up the weapons and stack the bodies over by the other building. I will keep the area cordoned off for a few minutes."

Dalton gathered a working party to move the bodies, and another with dirt to soak up the blood that had spilled on the area outside their compound. Within thirty minutes the area had been policed and the bodies were all well away from the compound. There were some on the back side of the compound that they missed, but with the large number of bodies one or two didn't make a lot of difference.

The soldiers spent the rest of the night trying to cover up any sign that they had been involved in the melee.

Clarence and Eddie spoke with the Chief of Police and smoothed out the tale a bit better. By the time the press finished with the questions of the civilian security guards at the compound next to where the battle took place, they had a good description of the force attacking the other people in the building, who were apparently into drugs based on the actions they had observed.

Meanwhile in Albuquerque, the same scenario had played out. The building was in an industrial area, although far enough away that no collateral damage was suffered there either. The outside troops had done most of the work there. They didn't want the assailants to get near the building or grounds, and at the first provocation, picked them off from behind and the left flank. There was no way to spin the carnage into anything other than what it was, an attack on the building in an attempt to steal the computer design, which was worth billions, and provided a lucrative target for industrial espionage, or downright piracy.

The force was much larger there and not much could be done to portray the incident any other way.

The police and emergency people were there into the next day trying to sort out and identify the dead and wounded. The police would question the wounded and get enough information to verify the building as the target, though they wouldn't get the real reason for the attack.

Eddie called Captain Leyland and told him what had happened. "I expect Carlos will be in your area within the next 24 hours. Did all the troops arrive and get settled in?"

"Yes sir. We're good to go. I can't even detect them from more than twelve feet away, and I know what to look for. There's no way Carlos will know we're here."

"Remember, I need the exact location where Carlos remains stationary for any length of time. Get cross references from other outposts so we can triangulate to determine the exact location of the portal."

"I already got that in place. When I contact them on the walkie talkie, they know to shoot the azimuth. We should be accurate to within three feet."

Beth and Cindy had remained in the office portion of the warehouse, and after the excitement had gone to the hotel, leaving Eddie and Clarence to deal with the aftermath. They were back by eight AM, and found Eddie and Clarence working on the oscillator.

"What do you want us to do?" Beth asked.

"Look at the plans and the ones in your head and see if you can figure out how the armature is held in place." He did not give them the benefit of his and Clarence's earlier discussion, wanting a new take on the problem.

After they had looked it over for several minutes and talked about it among themselves, Beth said to Eddie, "I think it is held in place by negative magnetism. The vacuum tube in the center, or whatever it is, will exert a pretty strong negative force, which will keep it in place equidistant from the core. I don't know how we get it set up to do that, but it appears that is the way it is supposed to be."

"I think it's about time to invite the feds to document this thing," Eddie said.

"I agree," Clarence added. "They might even be able to lend a hand in interpreting some of the more esoteric abbreviations for us."

Eddie called General Clarkson. "It appears we were not paranoid after all. I am going to send the extra troops back to you. It is now time for the technical people. Like I said, four is about the optimum, but six could probably be helpful if they are from different scientific disciplines."

"You want Chemists, Mathematicians, Engineers, and what else?"

"Someone in optics might come in handy, and one of your best people in the nuclear/radiation area."

"Why the last?" asked General Clarkson.

"Because I want him to estimate the effects of the output with relation to a safe zone away from the target area. I'm just going by intuition here, but we need to be able to estimate the destructive force of the system before we use it."

"That makes sense. I apologize for being short with you the last time we talked. I know you are under just as much pressure as I am, and you are on the scene and I am not. Just continue to let me know what you need and I will provide it."

"I think the need for most of the troops both here and in Albuquerque is at an end. It might help if you could spin the situation in Albuquerque in some manner as to identify a threat that had us worried and when we came to you the decision was made to provide personnel support due to the potential of the computer. It has been on the market long enough that people will certainly understand that."

"You're right. I will have my people put out a press release with some cockamamie story. It will have enough truth to be plausible. Have you briefed the President?"

"I am about to give him a call right now. I'm sure he has probably heard something about this by now, so I will ease his mind somewhat."

Eddie called and the President said, "I was wondering when I would hear from you. I have been getting all kinds of conflicting stories about both locations. What's the real story?"

"That's what has taken so long to get back to you. I didn't want to interrupt your sleep for no good reason, and I had to come up with a plausible tale for the press in Carson City. The Chief of Police has been very cooperative, though he doesn't know beans about what we are really doing. A nice commendation from you, nebulous of course, will thrill him and assure his continued cooperation. The story we decided on is that two drug factions had a shootout in the area next door to us. All the bodies were found near the other building, and there were enough bullet holes to make it plausible. The press even interviewed some of our civilian guards to get their impressions of what was happening next door. The guards were suspicious of the activity over there of course, so it appears that we are home free here."

"I talked with General Clarkson and he is going to put out a press release saying that a threat had been levied toward my company, which makes the innovative computer, and the government considered it credible and sent some troops to look into it. The ensuing battle occurred when the troops got close to the truth. There's enough truth in the statement to be plausible. He's going to send the technical people to us now for the oscillator and we will work on it full time until we get it ready."

"Is there anything else I can do?"

"I don't think so. I expect Carlos to make a trip to his portal, and Captain Leyland is prepared to get a precise location on it. Hopefully we can make some progress on the oscillator now. I know for sure we are going to have to visit Tom again, but I want to wait until we have more than one issue to deal with."

"Well keep at it, and keep me posted at least daily."

"I suppose by now you have some ideas kicking around in your head that you would like to talk about."

"Boy, were you ever right. I am having to work consciously to dial it back enough to appear normal."

Eddie laughed. "Wait until you get some briefing that nobody wants to make a call on and you dissect all the excuses as the briefing is going on. That's when you are going to have trouble controlling yourself."

"Go do your thing. I have taken enough of your time, but remember, at least daily."

Eddie hung up the phone. He now had to decide what to do about Carlos. He was confident that Captain Leyland could get a precise enough location if Carlos visited the portal, so Carlos had to be dealt with permanently.

He called Captain Leyland back. "We were talking about Carlos coming for a visit soon. If he does, and you are sure about the portal location, then have someone kill him, but make sure he is well away from the site. Not to tell you how to do your job, but an ambush of his car on the way out should work. Make sure you have enough people to do the job. One manifestation of his visit with the alien is enhanced physical powers, so don't underestimate him. If you have claymores then one or more in the road between the time he arrives and leaves might be prudent."

"I get the picture. He is not to escape under any circumstances."

"That's it."

"I will pull three of the teams off the least susceptible area and have them take care of it. We do in fact have claymores, and silenced sniper rifles. The only problem with the sniper rifle is finding a perch high enough to get a sight picture for the shooter."

"Find the highest spot you can and park one of the cars on the high spot. Then have the shooter lie on the roof to get his elevation."

"Gosh, you're so smart you could have even qualified for Special Forces," the Captain said playfully.

"Too many restrictions. I'm sure you can handle it. I'm just showing my ignorance of your training. My apologies. I expect a call within 24 hours saying that you have located the portal, and Carlos is no longer among the living."

"You got it."

Eddie next brought everyone up to date and they got back to work on the oscillator.

Chapter 30

The technical people arrived the following day. Eddie had expected them to be military, but there was only one military man in the group, and it was not the one he had the run in with previously. All were dressed in civilian clothing and as the introductions were made Eddie laid out the game plan.

"We have been gathering the material we think we will need in the quantities we envision using. The problem is that this item had never been built, and we are not sure about some of it. We all understand the design concept, and most of what goes into it, but we are not technically trained, and some of our methods will be trial and error. If you gentlemen see some way to do things more efficiently, please let us know. Also, as you go through the plans, if you run across a chemical element that you are not familiar with, please make note of it so we can resolve the problem."

"What do you mean an element we have not heard of?" one of the civilians asked.

"Are you absolutely sure that you are familiar with every chemical element in existence?" Eddie asked in return.

"Well no, but if you have discovered new elements then that fact is not known to the scientific community."

"Doctor, we are building something that is out of this world. Nobody has ever even conceived of it. If some of the technology seems strange or unfamiliar, give us the benefit of the doubt and make sure you take good notes so you can duplicate the process in the future."

Eddie continued. "The first dilemma for you gentlemen is to tell us how you would keep this armature," he said, pointing to the one almost completed that they had been working on, "In place without any visible means to attach it to anything else dealing with the design?"

"Here are the plans if you want to have a look."

All of them gathered around the drafting table where the plans had been laid out.

After studying the design for several minutes, one of them said, "I see no way to attach it."

"I didn't say attach it, I said keep it in place."

The group looked some more, but no one had any suggestions.

"The negative magnetic force exerted by the vacuum core will hold it in place, but we need to determine a method of getting the two mated without any outside influences affecting the set-up"

The scientists were enthralled with the design and spent the better part of the afternoon studying the design. Eddie and the others retired to a corner and discussed the vacuum arrangement. It was going to take a specialized lab to create the vacuum and get the required elements into the tube. They determined that they would need to consult with Tom about this as well.

Rather than everyone going back east again, it was decided that Eddie and Clarence would make the trip and the rest of them would continue working on the oscillator.

The two left Reno the following day and had landed in Philadelphia when Eddie got a call from Captain Leyland.

"Looks like your boy is back. I am watching him right now, and he is simply standing near the location where the tire tracks terminated in the clearing. I believe this is the location of the portal."

"Make sure you get accurate measurements from your other people. Have you made arrangements to take care of him when he leaves?"

"The people are already waiting for him. I called them on the walkie talkie when he arrived and had them plant the claymores. The marksman is set to go and five backup troops are on the scene. I don't see any possibility that he can get by us."

"Get the location, and call me when he has been disposed of. The body you can just bury in the scrub."

After he hung up he told Clarence, "That was Captain Leyland. Carlos has shown up at the location where we think the portal is located. They are going to get a precise location and then take care of Carlos."

"Are you sure you want to do that before you are absolutely sure about the portal?"

"I believe so. If the portal is not there when we get ready to take care of it, then the alien will have to wait for someone else to come along, and with the desolation of the location, it might be a long time before that happens. Since Carlos is the only link, then getting rid of him should solve the problem."

Meanwhile, back in Texas, Carlos was returning to his car and Leyland notified his people that he was coming and to take care of him.

The marksman was in a good location and when the car came into view, he lined up the shot. He was firing almost straight on to the car, and when he squeezed the trigger he knew the shot was accurate. The car did not stop when the bullet impacted, but kept rolling, and the claymores were detonated as the car triggered them.

The detail of men waited for a good five minutes before going to check the situation out. The marksman had done his job well, and Carlos's head was a bloody mass. The combination of the bullet and the claymore's had pretty much destroyed the car. Strangely enough, the engine was still running.

They were going to have to get rid of the car, so one of the men pushed Carlos's body aside and climbed into the car. He drove it off the road for about two hundred yards on the rims until he got to a location where they could partially bury the car, and disguise the location.

The leader of the detail called Captain Leyland and reported that the deed was done and told him what he had done with the car.

"I will send some additional men and I want you guys to dig a good sized hole for the car. Take the body someplace else and bury it."

He then called Eddie and reported that all was well.

Eddie called the President and gave him the information from the car on the way to Tom's location.

When they arrived, Eddie put the headpiece on and asked Tom about the problem areas they had identified. The armature problem was as they had thought. The negative magnetic force would hold it in place. The brackets that were welded to the housing was the only point at which the design would be joined, and that was simply to create stability within the housing.

The vacuum tube did not contain any elements they did not know about, but used combinations that humans would not have thought about. Eddie concentrated to make sure he understood all about that aspect, because the vacuum tubes would have to be fabricated in a physics lab someplace else.

Eddie told Tom about the other portal and what had happened. Tom was pleased that he had been able to locate it. He also told him that Joe had apparently activated whatever he had Eddie build into the first computer, and how it had affected Jesus.

"That is about what I thought would happen. You need to concentrate on the oscillator now to get rid of the portals."

"We will all be working on that until it is ready to go. Can you tell me how to test the system to make sure it works?"

"If you set them up equidistant from a given point and activate the system at its lowest setting, you will launch the object into space almost instantaneously. One second it will be there, and the next second it will be gone. If one did not know what was happening, it would simply appear that the object disappeared."

"The only power source is the batteries?" Eddie asked.

"The unit does not take a lot of power."

"Then I guess we will head back to work," Eddie said.

The group worked on the unit almost solid for the next two months. They had to make four more trips to talk to Tom about things they didn't understand, and the people the government sent to monitor the effort were completely baffled by the design. For every element Eddie and his people didn't understand, he made sure the scientists documented it in detail.

The hubbub had died down from the incidents at the two plants, and fortunately, nobody had put together the fact that the same company owned both sites.

Clarence was keeping in contact with the people in Yosemite, and had incorporated several of the military people into the surveillance team. The soldiers were to look for sites to locate the oscillator units when they were ready. They were given the exact parameters and used sextants to site the locations.

The first unit was finished in three months, a solid four weeks spent at the physics lab building the vacuum tubes. They decided to build all four, since they had the lab set up for that purpose and they would ultimately need four of them.

It was difficult to move the completed unit outside so they could get the next two pieces in. Though the unit was large and unwieldy, it did not exceed the maximum lift capacity of the heavy lift helicopters. The weight and moment of the completed design was calculated and lifting rings were welded to the outside of the apparatus.

Nothing out of the ordinary had happened at either of the sites they were watching, and six months passed before the units were ready. Eddie asked General Clarkson for four helicopters, and located a site in the desert that would suffice for the test. He called the

President and told him what was planned. "If you can get free to watch this, I can guarantee you will never forget it."

"When do you plan to do it?"

"As soon as everything is in place. I chose a desolate area rather than a government site to keep a low profile. I don't believe anyone will be able to discern what is happening, even if they get a glimpse of the results."

"Let me know where and when, so I can put together an official trip to someplace nearby. I really would like to witness this."

Eddie located a place in the Nevada desert near what the world knew as Area 51. It was government property, but was used for a munitions range and would suit their purposes.

General Clarkson notified the area commander that Eddie would be in touch and that he was to do whatever Eddie wanted.

The units were flown to the site and placed in the manner prescribed. They then placed a surplus government jeep at the intersection point. A sandbag enclosure was erected around the command post. The President had scheduled a trip to the west coast and made a precautionary landing at the nearest military facility, which was less than forty miles away.

When the President arrived everything was set up and ready to go. Eddie had wired the batteries to a single control unit at the lowest power setting as instructed. When all was in readiness he pressed the firing switch. Nothing seemed to happen, with the exception of the fact that the jeep was no longer visible.

The scientists had set up monitoring equipment to measure magnetic, electrical, nuclear, and kinetic energy generated by the unit.

The results were highly spiked readings of electrical and magnetic energy, with no nuclear and very little kinetic energy recorded. It looked like the unit operated as advertised.

The scientists had not been briefed on what would happen when the unit was activated and wanted to know where the jeep had disappeared to.

Eddie grinned at the President. "I guess the closest we can come to a good definition, is that the jeep was returned to its base elements. Either that or some invisible force whisked it away."

"What happens when that thing is used at full power in its current configuration?" the President asked.

"I think you will want each component to be maybe a mile away from the focal point, but at full power it will generate an electromagnetic pulse powerful enough to fry anything within its sphere of influence. We will get a better idea of the readings when we use it a couple of more times."

"Are you ready to do that now?"

"We are close. I want to go into Yosemite and site the locations before we fly the units up there. Because of the terrain, we are going to have a tough time with the aiming process to get the beams intersecting at the precise location we want them to. It may take some digging to get them set up. I have had a team working on that for the past month, and hopefully they are nearly done."

The President left and Eddie and the scientists reviewed the collected data again. If the system had done this with the lowest setting, Eddie extrapolated the readings that would be generated on full power if the increase was the same as the power increase. The power generated would be unbelievable.

The system components were rigged for travel and the helicopters flew them back to Carson City.

Eddie and Bill flew into Yosemite in a helicopter and had one of the people on site meet them a distance away from the portal. They were driven to the area and surveyed the situation.

The area's the troops had chosen to position the oscillators was about two hundred yards from the portal.

Eddie and Bill circled the area on foot and used an infrared measuring device to double check the locations. All were within a few feet of optimum and they used spray paint on the ground to show where the revised positions needed to be. The helicopters that would fly them in was going to be very close to detection distance from the portal if they had to maneuver due to wind currents.

Eddie explained in more detail to the work party what had to be done. "The trees in front of the location will have to be cut to give a clear path to the target area. Keep the area around the location clear of debris for the helicopter."

He also talked to the man leading the civilian crew that had been there from the start. "I don't see any reason to have your people stay any longer. There are enough troops here to keep the area secure now."

The man replied, "We would just as soon stay for the whole show if you don't mind. A couple of the guys have speculated that this whole drill has to do with aliens, and have about convinced the rest of us that what we have seen means that your project has a high level of interest from the government. Also the fact that we are guarding an empty space in the middle of nowhere is strange."

"I guess I can't deny you that, since you have been here for so long. I will warn you ahead of time though, that you will not witness anything spectacular. We have tested the system in the desert and it made a jeep disappear, but other than that there was no effects."

"By disappear, you don't mean destroyed, you mean it actually disappeared?"

"Right. It may have just broken up into molecules, but it was there one second, and the next second it was gone."

"Didn't you know what the system was going to do before you designed it?"

"Only in general terms. The system generates a huge amount of electromagnetic force, and the results of such force has never been witnessed, so there are a lot of unknowns. That was why we tested it in the desert. You will witness something that no one has ever seen, but you won't be able to talk about it afterwards."

"We'd like to stay all the same if you don't mind?"

"Then you are in. You might be able to give the troops a hand. I am thinking about having a grader flown in to get the slopes right."

He got on his phone right then and called Major Dalton. "I am going to need a road grader up here. Have Clarence get one and make arrangements for the chopper to fly it up. Send Clarence with it since he knows where we are located and tell him to have the chopper drop it at least a half mile from the location. He can contact me by phone and I will send someone to get it where the chopper drops it."

Eddie then called the President. "We are only a couple of days away, and I think it is about time to look at establishing a no fly zone over this area. I don't know how to go about that, but I figure you have people who do."

President Kelly said, "Can you give me a window?"

"We are finishing up the sites for the units now. I have a grader being flown in, which will speed up the process. We should be able to fly the units in tomorrow and the calibration process will not take all that long. If you figure day after tomorrow we should be ready. Have your people identify the time when there is less traffic in the area and we will try to schedule it for that time frame."

"I will get back to you later today with the particulars," he said.

The grader arrived later in the day and one of the army troops could operate it. With Bill and Eddie providing the supervision, they had all four of the needed locations ready.

The next day the units were flown in and positioned in a manner to focus their energy at the point designated.

Eddie had all the lines run from each unit to a location he had chosen as the command post.

The FAA had given the President a time close to eleven o'clock at night as the optimum time to close the area down to air traffic. The FAA controllers were notified of the airspace closure and would report to the Military base near Las Vegas, who would verify that the area was clear to General Clark, who would call Eddie with the information.

All the instrumentation was set up beyond the oscillator units, and when the time arrived Eddie warned everyone and energized the units.

What happened was so spectacular that it almost defied description. A sudden very bright and multicolored circle of light erupted from the clearing where the mechanism was focused. The design was so symmetrical that it looked like a directed beam and was straight up from the location. The phenomenon lasted for about three seconds and winked out as suddenly as it had appeared.

Eddie asked everyone else to stay put, and he and Bill walked to the clearing where the portal had been. As they entered the clearing, neither of them experienced the presence of Joe. "Well," Eddie said, "It looks like it worked as advertised."

Eddie next pulled out his phone and called the President, who was awaiting the call, though it was two o'clock in the morning on the east coast. "The media is going to be inundated with phone calls about a strange light in the sky over Yosemite. It was absolutely awesome. The portal was so symmetrical the light went straight up with no light spilling to the sides."

"I wish we had thought to film it," he said.

"Maybe we want to do that with the next one. I haven't checked the readings from the scientists yet, so I don't know if we will be able to quantify the energy output, but believe me, it far surpasses anything we have now."

"I guess it's on to Texas now?"

"We will start moving tomorrow."

"Keep in touch."

Chapter 31

The units were flown to a railhead by helicopter and loaded aboard flatcars. The train then headed for Texas, the trip taking two days because of the roundabout routing necessary to get them to the desired location.

Eddie and his people flew on the business jet and spent the next day looking the location over and double checking the location of the portal. The land was flat enough that they would not have to do much to position the units in the proper places.

The newspapers on that day were full of stories of the bright unexplained light in the night sky over California. Someone had luckily gotten a picture of the phenomenon and that had been posted on the internet. The hits were so many that the servers couldn't keep up with the traffic. Fortunately, nobody had an explanation, but speculation ran the gamut from extra terrestrials to some secret government weapon.

The airspace closure over the area at the time of the sighting came to light and various government agencies fielded calls, which were answered truthfully for the most part. "We simply don't know," was the general response.

They did the same thing for the operation in Texas, except this time they enlisted a professional filming crew to record the proceedings. They were not told what was in the clearing where the beams were focused, and hopefully the film would not show anything but the bright light as had happened in California. The story was that it was a new weapon the government was testing for missile defense of the nation.

The airspace closures went into effect at midnight and the process was repeated. Even knowing what was coming, the actuality of the event was just as spectacular as the first time.

Eddie, Bill and Clarence checked the area to be sure the portal was no longer there. It wasn't.

Again Eddie called the President with the result of the operation. "Things went like before. We got the film and I will bring it to you very soon. We are going to take the units back to Carson City until we decide where to go from here. Besides which we need to pay another visit to Tom."

"Get by to see me as soon as you can. I want to see the event on film. Wait a minute, I have the television on and there's a breaking news story. I think it might be about your little activity there." He turned the sound up.

As Eddie waited he could hear the television commentator in the background. He was talking about a phenomenon similar to what had happened a couple of days before over California. This time a video of the event accompanied the story. The cameraman must have been a long distance away, because the light was far away and looked like a lightning bolt, but with much greater definition. It was also a direct beam, as straight as an arrow as the saying goes.

After the President saw the pictures on television, he said to Eddie, "some newsman had his camera pointed in the right direction at the proper time and got a view of it from a distance. Even that was very impressive."

"I am wondering what the presentation will be like when there is no portal to contain the energy?" Eddie said.

"That's something we are going to have to find out."

"All the more reason to talk to Tom at the earliest opportunity."

Eddie, Bill and Clarence left for Washington the following morning. Captain Leyland was told to sanitize the area and send the troops back. Eddie got a copy of the film on a DVD and told the camera crew not to release the imagery to anyone until he gave them the okay.

When they got in to see the President, he had a DVD player set up in the Oval Office. Eddie put the DVD in and played it. The speed of the event was such that even at the slowest speed, they could not tell exactly when the energy started. The camera tried to follow the beam upward, but by the time he got the camera to that angle, the light was already such that you could not tell the beginning and end, but it was as symmetrical as the first had been.

The President said, "I wonder if the light ceases when it leaves the atmosphere, or continues on?"

Eddie answered, "I believe it probably diminishes. Remember how far away the target really is."

"I hadn't considered that, but it would take it a thousand years to reach the destination traveling at the speed of light."

"Exactly, so what I think happens is that it destroys the portal from the upper atmosphere to the earth and possibly affects the precision of the beam from there to the ship."

"Either way, the portals are gone for now."

"We are going to see Tom today. I hope to get some answers about how to use the system for EMP and how it generates electric power."

"Good luck with that. It will go a long way to placating the public if we show them a way to generate energy and save money in the process."

They left the White House and flew to Philadelphia. A rented car got them to the location and the three of them walked to the grove of trees.

The contact came as always then Tom said, "Since you are here, I assume the device worked."

"It did," Eddie answered. "Joe and Harry are no longer with us."

"I am happy for you. But they are going to know that someone interfered with their plans. I think it might be prudent for you to destroy my portal as well. That will avert their suspicions to someone else whose portal is still in place."

"That will mean that you have to build another for yourself," Eddie said.

"Yes, but time is not the same to us as it is to your kind."

"How many earth years would it take you to build another?" Bill inquired.

"About two, more or less."

"Do you choose the location where it terminates, or is it random?"

"We have used random locations in the past, but I think I can have it terminate in a location of my choice, now that I know more about your planet."

"If we give you a specific location, can you have it terminate there?"

"I believe so. You are thinking that you will contact me when I return?"

"Exactly. We can prepare a place that is private and control access to the area. We certainly have the money to purchase property someplace and set up for that."

"Identify a place for me and I will attempt to do that."

"We need to get the equipment in place, so we will contact you again later. How do we use the device as an electric generator?"

"Build the electric generator as if it were powered by electricity and wire it to one of the oscillators. You can adjust the power setting for the output to control how much electricity you want to generate."

"What happens to the magnetic output?" Eddie asked.

"You can focus that to power a larger generator nearby."

"One other question," Eddie asked, "How long will these units last?"

"Until the copper corrodes, or some metal component fatigues enough to break. I should think twenty years at a minimum."

"By then you should be back to show us how to fix it. And the EMP application? How does that work?"

"Use the four units and position them in such a way that when the output meets it does so at an angle. The angle of the beams skews the output in a cone shape and with distance the cone widens. A one degree angle at the surface would expand to a circular area of fifty miles before the energy leaves your atmosphere."

"What happens to the energy once it leaves our atmosphere?"

"It disperses because there is nothing in space to channel the energy. Eventually it just ceases to exist as energy and returns to its base elements."

"We will contact you again when everything is ready here, and give you the coordinates for your next portal at the same time."

They departed and on the way to Philadelphia, discussed a new location for the portal. "Do we want to tell the President about this new portal?"

Bill answered, "I don't think so. If at some later time we see a need, we can contact him and tell him then."

"I agree," Clarence said. "You probably shouldn't even tell me."

"You have been part of this too long to be left out. We will try to find a location that is convenient for you. I am thinking that we might move our company headquarters and plan for the portal in a courtyard or something like that," Eddie said.

"That's a great idea. Since the factory in Georgia is so successful, why not look for something close to that?" Bill asked.

"We had better not plan anything like this without some input from the girls," Eddie observed.

"You're right about that. I wasn't thinking there for a minute," Bill replied.

"You guys are going to have to buy your own jet now and let me have mine back," Clarence observed.

"We will do that," Eddie said. "I think we should keep Captain Leyland around too. He knows about the aliens, and will make a good man to run security for us, or anything else for that matter."

"I also agree with that. He can be made aware of what the plan is for the new portal and see that nobody messes that up."

The group stopped back in Washington and had a quick visit with the President, where Eddie told him what Tom had said about the power generation and the EMP use. "I know it's a bit early yet, but do you think you might run for a second term?"

"I have been thinking a lot about that lately, and I have decided that you are right. The job would be too restrictive, and it will be much harder to keep this a secret in this job. I have a person in mind to start lobbying for the job, but I haven't talked to him about it yet."

"Who do you have in mind?" Bill asked.

"Strangely enough, Clarence. He has been in contact for more than forty years and has managed to keep it secret, and he certainly is smart enough for the job. He is independently wealthy, and has no bad press. The only blemish on his record would be his teenage years, and that was so long ago that nobody would care, even if something did come out, which it probably won't."

"His involvement in development of the power generator will boost his stock, and the fact that he did it without government funds makes it that much better. I think I can get the nomination for him in the eighteen months I have left in office."

"What do you think Clarence? Would you like to run the country and try to fix what's broken?" Eddie asked.

"What I really think is that you would be better for the job than me."

The President said, "It's something to think about. We will discuss it in more depth later"

"Another thing we haven't told you is that Tom thinks it would be better if we destroyed his portal too, to shift suspicion away from him. It is not likely that the other two will think he had anything to do with it if his is destroyed too."

"That's very generous of him, and I think he might be right about the suspicion being shifted away from himself."

"We are going to make arrangements to have the entire unit shipped back here. After we destroy Tom's portal we can turn the unit over to you. You need to be thinking about a place to base it, and who's going to run the program. I think the scientists know enough now to operate the thing, and make any necessary repairs. We will still be around to assist if we are needed."

"General Clarkson needs to be briefed. Would you mind doing that?" Eddie asked.

"I will invite him over for lunch or something and have a long talk with him."

The group flew back to Albuquerque and gave the girls a blow by blow of the meeting. "What do you think we should do about giving Tom a location for his next portal?" Bill asked.

Beth and Cindy looked at each other. Finally Beth said, "We both kind of like it here, and with the building so new and working out so well, why not just stay here. We can buy the adjacent property for expansion and when we build, incorporate an open courtyard in the center of the complex. That would really be ideal."

"Okay, that's settled. We stay here, but someone is going to have to check into the property rather quickly so we can get the coordinates to Tom before we demolish his portal."

"I will take care of that just as soon as I can," Beth said.

The others started making arrangements for the units to be moved back east. Eddie also contacted Captain Leyland and asked him if he would be willing to resign his commission in the army to come to work for them.

"Does this have to do with the same situation?" he asked.

"I need to talk to you in person to lay it all out, but your salary will be five times what you make in the military with a lot of perks. We will take care of you financially."

"I am definitely interested, but I don't know how easy it will be to resign from the army."

"I have friends in high places, as you may recall, so I don't see that as an impediment."

"Then count me in."

Now that the decision to obliterate Tom's portal was made they had to arrange for shipment of the units to Pennsylvania. Clarence told them he would work on that aspect of the problem.

By the following day, Beth had identified the owner of the adjacent property and made them an offer they couldn't refuse. Eddie, Beth, Bill and Cindy walked the property and tried to visualize what they would build on it.

Beth said, "What I visualize is a coliseum like design, with the outer stories of the building being higher than the inner ones. The architect can determine how many stories the outside will have and reduce the height by one story until he gets near the center point. We will leave a quadrangle in the center, with the portal location right in the middle and enclosed by a fence, or statuary, or something like that. It will provide a good place for recreation for the employees and add to the aesthetics at the same time."

"That works for me," Eddie said. "We need to determine the center point so I can give Tom the coordinates."

"I will have a surveyor stake the property tomorrow and you can use GPS to get the location."

"Can anyone else think of anything we have missed?"

"What about the property in Carson City?" Bill asked.

"It would be nice to do something for the city, since they were so cooperative," Cindy said. "Why not tear the building down and make it into a nice park. It will provide a good place for kids to play, and there is not much else in that neighborhood."

"That also is a good idea, will you check into that?" Eddie asked.

Bill said, "Some of the more senior people we have worked with know that something about us is not what it should be, and I think we should make an effort to keep them on the payroll in some capacity. The guys in Yosemite know too much to just turn them loose, and if we take care of them, they will have less reason to try to exploit the situation. I have a feeling we are going to expand into a lot of other areas, and they may have some good ideas."

"You mean something like I did with Captain Leyland?"

"Exactly. We can spread them around to areas that they are interested in, and let them oversee those activities. The computer thing was enough to show me that the four of us are going to be spread very thin if we go in many more directions."

"That sounds good to me," Cindy said.

"Okay, all the guys in Yosemite, Major Dalton, if he is willing, and a couple more from Carson City. We will tell Leyland to bring along three or four who were more involved in the thing in Texas. Maybe we should tell him to bring the ones involved in taking care of Carlos," Eddie said.

"Who else," Cindy asked.

"Our staff here stays, and the Chief of Security definitely retains his position."

"Are we getting too ambitious here?" Eddie asked.

"I don't think so. We have made nearly three billion on the computer already, and there is the potential for a hundred billion over time," Cindy said.

"Okay, let's run with that. Bill and I will take care of the portal in Pennsylvania and be back here after that."

Chapter 32

It took three days to get the units to the area where they wanted them, and when they were ready, Bill and Eddie had a last visit with Tom.

"We're about ready with everything, so if there are any last minute instructions, now is the time," Eddie said to Tom

"I only want to say that I am proud of you for the way you handled everything, and I look forward to seeing you again in the future. Until then I wish you the best."

They did the airspace closure for the area, which was much more difficult for the busy eastern seaboard. This time they allowed a network cameraman to film the event, though it was billed as a test of an EMP weapon rather than what it really was. With the live feed, the entire world knew about the event in real time.

The President came on after the event and explained to the public that what they had witnessed was the test of a weapon the government in concert with private industry had been working on for quite some time. He went on to explain that there were weapons which were designed to cripple an area by destroying all their mechanical and electrical devices. The new EMP generator that had been developed was a weapon to defend against such an attack. And oh, by the way, the new weapon could produce enough electrical power to light a small city from a simple twelve volt battery, and that technology would be pursued in the future.

The event increased the President's popularity and he used the opportunity to introduce Clarence to the public as a private partner in the recent developments. Clarence was open and intelligent and made a good subject for the political pundits. As time went on, the President exerted more pressure on his party to look seriously at Clarence as a possible replacement for him on the next ticket. He intimated that his health was not as good as it could be and that he was considering withdrawing from the next presidential campaign.

In the meantime, the building in Albuquerque was well underway, and the design was innovative enough that many architects were pursuing similar projects, even before theirs was done.

The group had integrated the new people into their operations and everyone was content with their lot.

Less than two years later, the building was dedicated and they were deciding what to put into the massive amount of space they now had available to them. The computer was still selling steadily and the coffers were growing, even with the additional costs involved with the new project.

Carson City, Nevada now had an upscale park and playground with all the modern things kids liked to play on, and the hotel had been converted to a shelter and food kitchen for the needy. The company staffed the place and paid all the expenses on a continuing basis.

The President let it slip that Clarence Woodman had also been involved in that endeavor, and his stock rose enough to get the nomination from the party in the Presidential primary.

When the elections were held, Clarence won by a hefty margin, though it was not a complete runaway. The President had tutored him in some of the intricacies of the office, and by the time January 20th rolled around, he was an old hand.

The group from Albuquerque attended the inauguration as invited guests and were in touch with Clarence at least once a week.

The group kept a close check on the portal location, and one day some two and a half years later, Beth had gone alone to check and to her surprise Tom made contact with her. She quickly brought the other three down and they all had a reunion.

Eddie said, "I want to know what happened with Joe and Harry?"

Tom laughed, "It was beautiful. When I arrived back at the ship, I got on the communications channel with them and everyone wondered what had happened. It seems that they have a theory that some of the others have banded against the three of us. We all are back now, though I have no idea where the other two ended up. You will have to keep an eye out for the signs on new contact with them. Now what has been happening since I left?"

"Well, our President decided not to run for reelection and put Clarence in the limelight enough to get the nomination. Clarence is now our new President."

"I am glad for that. Clarence was always faithful to report to me, but I sensed that he was a lonely man. Your befriending him was a real blessing to him, and I am happy that he is now in a position to do some real good."

"You are now in a quadrangle inside a stadium like structure that we had built. Access will be strictly controlled, so if you want new company, let us know. There is one gentleman I would like you to meet. He believed in you guys even before he knew we had made contact, and was very essential to taking care of Harry in Texas. I don't think you should increase his awareness, but I just want him to know what his dreams have been about."

"The former President is out in the world on his own now, and making more money than you can count. He is also doing a lot of good and you would be proud of him."

"I am proud of all of you."

"We will probably be visiting you a lot more often now, since you are so convenient."

"I would like that, and just so you know, I think I am now the unofficial leader of the pack."

www.ingramcontent.com/pod-product-compliance
Lightning Source LLC
Chambersburg PA
CBHW062123170626
46813CB00002B/551